T0212925

Lecture Notes in Computer Science 14565

The series Lecture Notes in Computer Science (LNCS), including its subseries Lecture Notes in Artificial Intelligence (LNAI) and Lecture Notes in Bioinformatics (LNBI), has established itself as a medium for the publication of new developments in computer science and information technology research, teaching, and education.

LNCS enjoys close cooperation with the computer science R & D community, the series counts many renowned academics among its volume editors and paper authors, and collaborates with prestigious societies. Its mission is to serve this international community by providing an invaluable service, mainly focused on the publication of conference and workshop proceedings and postproceedings. LNCS commenced publication in 1973.

Stevan Rudinac · Alan Hanjalic · Cynthia Liem ·
Marcel Worring · Björn Þór Jónsson · Bei Liu ·
Yoko Yamakata
Editors

MultiMedia Modeling

30th International Conference, MMM 2024
Amsterdam, The Netherlands, January 29 – February 2, 2024
Proceedings, Part V

 Springer

Editors
Stevan Rudinac 🆔
University of Amsterdam
Amsterdam, The Netherlands

Alan Hanjalic 🆔
Delft University of Technology
Delft, The Netherlands

Cynthia Liem 🆔
Delft University of Technology
Delft, The Netherlands

Marcel Worring 🆔
University of Amsterdam
Amsterdam, The Netherlands

Björn Þór Jónsson 🆔
Reykjavik University
Reykjavik, Iceland

Bei Liu 🆔
Microsoft Research Lab – Asia
Beijing, China

Yoko Yamakata 🆔
The University of Tokyo
Tokyo, Japan

ISSN 0302-9743   ISSN 1611-3349 (electronic)
Lecture Notes in Computer Science
ISBN 978-3-031-56434-5 ISBN 978-3-031-56435-2 (eBook)
https://doi.org/10.1007/978-3-031-56435-2

This Springer imprint is published by the registered company Springer Nature Switzerland AG
The registered company address is: Gewerbestrasse 11, 6330 Cham, Switzerland

Paper in this product is recyclable.

Preface

These five-volume proceedings contain the papers presented at MMM 2024, the International Conference on Multimedia Modeling. This 30th anniversary edition of the conference was held in Amsterdam, The Netherlands, from 29 January to 2 February 2024. The event showcased recent research developments in a broad spectrum of topics related to multimedia modelling, particularly: audio, image, video processing, coding and compression, multimodal analysis for retrieval applications, and multimedia fusion methods.

We received 297 regular, special session, Brave New Ideas, demonstration and Video Browser Showdown paper submissions. Out of 238 submitted regular papers, 27 were selected for oral and 86 for poster presentation through a double-blind review process in which, on average, each paper was judged by at least three program committee members and reviewers. In addition, the conference featured 23 special session papers, 2 Brave New Ideas and 8 demonstrations. The following four special sessions were part of the MMM 2024 program:

- FMM: Special Session on Foundation Models for Multimedia
- MDRE: Special Session on Multimedia Datasets for Repeatable Experimentation
- ICDAR: Special Session on Intelligent Cross-Data Analysis and Retrieval
- XR-MACCI: Special Session on eXtended Reality and Multimedia - Advancing Content Creation and Interaction

The program further included four inspiring keynote talks by Anna Vilanova from the Eindhoven University of Technology, Cees Snoek from the University of Amsterdam, Fleur Zeldenrust from the Radboud University and Ioannis Kompatsiaris from CERTH-ITI.

In addition, the annual MediaEval workshop was organised in conjunction with the conference. The attractive and high-quality program was completed by the Video Browser Showdown, an annual live video retrieval competition, in which 13 teams participated.

We would like to thank the members of the organizing committee, special session and VBS organisers, steering and technical program committee members, reviewers, keynote speakers and authors for making MMM 2024 a success.

December 2023

Stevan Rudinac
Alan Hanjalic
Cynthia Liem
Marcel Worring
Björn Þór Jónsson
Bei Liu
Yoko Yamakata

Organization

General Chairs

Stevan Rudinac   University of Amsterdam, The Netherlands
Alan Hanjalic   Delft University of Technology, The Netherlands
Cynthia Liem   Delft University of Technology, The Netherlands
Marcel Worring   University of Amsterdam, The Netherlands

Technical Program Chairs

Björn Þór Jónsson   Reykjavik University, Iceland
Bei Liu   Microsoft Research, China
Yoko Yamakata   University of Tokyo, Japan

Community Direction Chairs

Lucia Vadicamo   ISTI-CNR, Italy
Ichiro Ide   Nagoya University, Japan
Vasileios Mezaris   Information Technologies Institute, Greece

Demo Chairs

Liting Zhou   Dublin City University, Ireland
Binh Nguyen   University of Science, Vietnam National University Ho Chi Minh City, Vietnam

Web Chairs

Nanne van Noord   University of Amsterdam, The Netherlands
Yen-Chia Hsu   University of Amsterdam, The Netherlands

Video Browser Showdown Organization Committee

Klaus Schoeffmann	Klagenfurt University, Austria
Werner Bailer	Joanneum Research, Austria
Jakub Lokoc	Charles University in Prague, Czech Republic
Cathal Gurrin	Dublin City University, Ireland
Luca Rossetto	University of Zurich, Switzerland

MediaEval Liaison

Martha Larson	Radboud University, The Netherlands

MMM Conference Liaison

Cathal Gurrin	Dublin City University, Ireland

Local Arrangements

Emily Gale	University of Amsterdam, The Netherlands

Steering Committee

Phoebe Chen	La Trobe University, Australia
Tat-Seng Chua	National University of Singapore, Singapore
Kiyoharu Aizawa	University of Tokyo, Japan
Cathal Gurrin	Dublin City University, Ireland
Benoit Huet	Eurecom, France
Klaus Schoeffmann	Klagenfurt University, Austria
Richang Hong	Hefei University of Technology, China
Björn Þór Jónsson	Reykjavik University, Iceland
Guo-Jun Qi	University of Central Florida, USA
Wen-Huang Cheng	National Chiao Tung University, Taiwan
Peng Cui	Tsinghua University, China
Duc-Tien Dang-Nguyen	University of Bergen, Norway

Special Session Organizers

FMM: Special Session on Foundation Models for Multimedia

Xirong Li	Renmin University of China, China
Zhineng Chen	Fudan University, China
Xing Xu	University of Electronic Science and Technology of China, China
Symeon (Akis) Papadopoulos	Centre for Research and Technology Hellas, Greece
Jing Liu	Chinese Academy of Sciences, China

MDRE: Special Session on Multimedia Datasets for Repeatable Experimentation

Klaus Schöffmann	Klagenfurt University, Austria
Björn Þór Jónsson	Reykjavik University, Iceland
Cathal Gurrin	Dublin City University, Ireland
Duc-Tien Dang-Nguyen	University of Bergen, Norway
Liting Zhou	Dublin City University, Ireland

ICDAR: Special Session on Intelligent Cross-Data Analysis and Retrieval

Minh-Son Dao	National Institute of Information and Communications Technology, Japan
Michael Alexander Riegler	Simula Metropolitan Center for Digital Engineering, Norway
Duc-Tien Dang-Nguyen	University of Bergen, Norway
Binh Nguyen	University of Science, Vietnam National University Ho Chi Minh City, Vietnam

XR-MACCI: Special Session on eXtended Reality and Multimedia - Advancing Content Creation and Interaction

Claudio Gennaro	Information Science and Technologies Institute, National Research Council, Italy
Sotiris Diplaris	Information Technologies Institute, Centre for Research and Technology Hellas, Greece
Stefanos Vrochidis	Information Technologies Institute, Centre for Research and Technology Hellas, Greece
Heiko Schuldt	University of Basel, Switzerland
Werner Bailer	Joanneum Research, Austria

Program Committee

Alan Smeaton	Dublin City University, Ireland
Anh-Khoa Tran	National Institute of Information and Communications Technology, Japan
Chih-Wei Lin	Fujian Agriculture and Forestry University, China
Chutisant Kerdvibulvech	National Institute of Development Administration, Thailand
Cong-Thang Truong	Aizu University, Japan
Fan Zhang	Macau University of Science and Technology/Communication University of Zhejiang, China
Hilmil Pradana	Sepuluh Nopember Institute of Technology, Indonesia
Huy Quang Ung	KDDI Research, Inc., Japan
Jakub Lokoc	Charles University, Czech Republic
Jiyi Li	University of Yamanashi, Japan
Koichi Shinoda	Tokyo Institute of Technology, Japan
Konstantinos Ioannidis	Centre for Research & Technology Hellas/Information Technologies Institute, Greece
Kyoung-Sook Kim	National Institute of Advanced Industrial Science and Technology, Japan
Ladislav Peska	Charles University, Czech Republic
Li Yu	Huazhong University of Science and Technology, China
Linlin Shen	Shenzhen University, China
Luca Rossetto	University of Zurich, Switzerland
Maarten Michiel Sukel	University of Amsterdam, The Netherlands
Martin Winter	Joanneum Research, Austria
Naoko Nitta	Mukogawa Women's University, Japan
Naye Ji	Communication University of Zhejiang, China
Nhat-Minh Pham-Quang	Aimesoft JSC, Vietnam
Pierre-Etienne Martin	Max Planck Institute for Evolutionary Anthropology, Germany
Shaodong Li	Guangxi University, China
Sheng Li	National Institute of Information and Communications Technology, Japan
Stefanie Onsori-Wechtitsch	Joanneum Research, Austria
Takayuki Nakatsuka	National Institute of Advanced Industrial Science and Technology, Japan
Tao Peng	UT Southwestern Medical Center, USA

Thitirat Siriborvornratanakul	National Institute of Development Administration, Thailand
Vajira Thambawita	SimulaMet, Norway
Wei-Ta Chu	National Cheng Kung University, Taiwan
Wenbin Gan	National Institute of Information and Communications Technology, Japan
Xiangling Ding	Hunan University of Science and Technology, China
Xiao Luo	University of California, Los Angeles, USA
Xiaoshan Yang	Institute of Automation, Chinese Academy of Sciences, China
Xiaozhou Ye	AsiaInfo, China
Xu Wang	Shanghai Institute of Microsystem and Information Technology, China
Yasutomo Kawanishi	RIKEN, Japan
Yijia Zhang	Dalian Maritime University, China
Yuan Lin	Kristiania University College, Norway
Zhenhua Yu	Ningxia University, China
Weifeng Liu	China University of Petroleum, China

Additional Reviewers

Alberto Valese
Alexander Shvets
Ali Abdari
Bei Liu
Ben Liang
Benno Weck
Bo Wang
Bowen Wang
Carlo Bretti
Carlos Cancino-Chacón
Chen-Hsiu Huang
Chengjie Bai
Chenlin Zhao
Chenyang Lyu
Chi-Yu Chen
Chinmaya Laxmikant Kaundanya
Christos Koutlis
Chunyin Sheng
Dennis Hoppe
Dexu Yao
Die Yu

Dimitris Karageorgiou
Dong Zhang
Duy Dong Le
Evlampios Apostolidis
Fahong Wang
Fang Yang
Fanran Sun
Fazhi He
Feng Chen
Fengfa Li
Florian Spiess
Fuyang Yu
Gang Yang
Gopi Krishna Erabati
Graham Healy
Guangjie Yang
Guangrui Liu
Guangyu Gao
Guanming Liu
Guohua Lv
Guowei Wang

Gylfi Þór Guðmundsson
Hai Yang Zhang
Hannes Fassold
Hao Li
Hao-Yuan Ma
Haochen He
Haotian Wu
Haoyang Ma
Haozheng Zhang
Herng-Hua Chang
Honglei Zhang
Honglei Zheng
Hu Lu
Hua Chen
Hua Li Du
Huang Lipeng
Huanyu Mei
Huishan Yang
Ilias Koulalis
Ioannis Paraskevopoulos
Ioannis Sarridis
Javier Huertas-Tato
Jiacheng Zhang
Jiahuan Wang
Jianbo Xiong
Jiancheng Huang
Jiang Deng
Jiaqi Qiu
Jiashuang Zhou
Jiaxin Bai
Jiaxin Li
Jiayu Bao
Jie Lei
Jing Zhang
Jingjing Xie
Jixuan Hong
Jun Li
Jun Sang
Jun Wu
Jun-Cheng Chen
Juntao Huang
Junzhou Chen
Kai Wang
Kai-Uwe Barthel
Kang Yi

Kangkang Feng
Katashi Nagao
Kedi Qiu
Kha-Luan Pham
Khawla Ben Salah
Konstantin Schall
Konstantinos Apostolidis
Konstantinos Triaridis
Kun Zhang
Lantao Wang
Lei Wang
Li Yan
Liang Zhu
Ling Shengrong
Ling Xiao
Linyi Qian
Linzi Xing
Liting Zhou
Liu Junpeng
Liyun Xu
Loris Sauter
Lu Zhang
Luca Ciampi
Luca Rossetto
Luotao Zhang
Ly-Duyen Tran
Mario Taschwer
Marta Micheli
Masatoshi Hamanaka
Meiling Ning
Meng Jie Zhang
Meng Lin
Mengying Xu
Minh-Van Nguyen
Muyuan Liu
Naomi Ubina
Naushad Alam
Nicola Messina
Nima Yazdani
Omar Shahbaz Khan
Panagiotis Kasnesis
Pantid Chantangphol
Peide Zhu
Pingping Cai
Qian Cao

Qian Qiao
Qiang Chen
Qiulin Li
Qiuxian Li
Quoc-Huy Trinh
Rahel Arnold
Ralph Gasser
Ricardo Rios M. Do Carmo
Rim Afdhal
Ruichen Li
Ruilin Yao
Sahar Nasirihaghighi
Sanyi Zhang
Shahram Ghandeharizadeh
Shan Cao
Shaomin Xie
Shengbin Meng
Shengjia Zhang
Shihichi Ka
Shilong Yu
Shize Wang
Shuai Wang
Shuaiwei Wang
Shukai Liu
Shuo Wang
Shuxiang Song
Sizheng Guo
Song-Lu Chen
Songkang Dai
Songwei Pei
Stefanos Iordanis Papadopoulos
Stuart James
Su Chang Quan
Sze An Peter Tan
Takafumi Nakanishi
Tanya Koohpayeh Araghi
Tao Zhang
Theodor Clemens Wulff
Thu Nguyen
Tianxiang Zhao
Tianyou Chang
Tiaobo Ji
Ting Liu
Ting Peng
Tongwei Ma

Trung-Nghia Le
Ujjwal Sharma
Van-Tien Nguyen
Van-Tu Ninh
Vasilis Sitokonstantinou
Viet-Tham Huynh
Wang Sicheng
Wang Zhou
Wei Liu
Weilong Zhang
Wenjie Deng
Wenjie Wu
Wenjie Xing
Wenjun Gan
Wenlong Lu
Wenzhu Yang
Xi Xiao
Xiang Li
Xiangzheng Li
Xiaochen Yuan
Xiaohai Zhang
Xiaohui Liang
Xiaoming Mao
Xiaopei Hu
Xiaopeng Hu
Xiaoting Li
Xiaotong Bu
Xin Chen
Xin Dong
Xin Zhi
Xinyu Li
Xiran Zhang
Xitie Zhang
Xu Chen
Xuan-Nam Cao
Xueyang Qin
Xutong Cui
Xuyang Luo
Yan Gao
Yan Ke
Yanyan Jiao
Yao Zhang
Yaoqin Luo
Yehong Pan
Yi Jiang

Yi Rong

Yi Zhang

Yihang Zhou

Yinqing Cheng

Yinzhou Zhang

Yiru Zhang

Yizhi Luo

Yonghao Wan

Yongkang Ding

Yongliang Xu

Yosuke Tsuchiya

Youkai Wang

Yu Boan

Yuan Zhou

Yuanjian He

Yuanyuan Liu

Yuanyuan Xu

Yufeng Chen

Yuhang Yang

Yulong Wang

Yunzhou Jiang

Yuqi Li

Yuxuan Zhang

Zebin Li

Zhangziyi Zhou

Zhanjie Jin

Zhao Liu

Zhe Kong

Zhen Wang

Zheng Zhong

Zhengye Shen

Zhenlei Cui

Zhibin Zhang

Zhongjie Hu

Zhongliang Wang

Zijian Lin

Zimi Lv

Zituo Li

Zixuan Hong

Contents – Part V

RESET: Relational Similarity Extension for V3C1 Video Dataset

Patrik Veselý⬤ and Ladislav Peška[✉]⬤

Faculty of Mathematics and Physics, Charles University, Prague, Czechia
ladislav.peska@matfyz.cuni.cz

Abstract. Effective content-based information retrieval (IR) is crucial across multimedia platforms, especially in the realm of videos. Whether navigating a personal home video collection or browsing a vast streaming service like YouTube, users often find that a simple metadata search falls short of meeting their information needs. Achieving a reliable estimation of visual similarity holds paramount significance for various IR applications, such as query-by-example, results clustering, and relevance feedback. While many pre-trained models exist for this purpose, they often mismatch with human-perceived similarity leading to biased retrieval results. Up until now, the practicality of fine-tuning such models has been hindered by the absence of suitable datasets.

This paper introduces RESET: RElational Similarity Evaluation dataseT. RESET contains over 17,000 similarity annotations for query-candidate-candidate triples of video keyframes taken from the publicly available V3C1 video collection. RESET addresses both close and distant triplets within the realm of unconstrained V3C1 imagery and two of its compact sub-domains: wedding and diving. Offering fine-grained similarity annotations along with their context, re-annotations by multiple users, and similarity estimations from 30 pre-trained models, RESET serves dual purposes. It facilitates the evaluation of novel visual embedding models w.r.t. similarity preservation and provides a resource for fine-tuning visual embeddings to better align with human-perceived similarity. The dataset is available from https://osf.io/ruh5k.

Keywords: Visual similarity · content-based video retrieval · human annotations

1 Introduction

We live in a time where every cellphone can produce tons of multimedia daily. We regularly bring home thousands of holiday photos or film hours of videos recording precious moments of our life - or just for fun. Also, semi-professional multimedia production is being boosted by the emergence of video-on-demand (VOD) platforms such as YouTube or TikTok. As a result, the size of multimedia data grows perpetually with astronomical speed.

S. Rudinac et al. (Eds.): MMM 2024, LNCS 14565, pp. 1–14, 2024.
https://doi.org/10.1007/978-3-031-56435-2_1

However, the ability to retrieve relevant information often lags behind the data production. The VOD search engines still predominantly utilize items metadata or simple relations between them (e.g., subscriptions). This, however, may not be enough to cover some of the user's information needs, so a tedious sequential search is often the only option left.

Nonetheless, methods to mitigate the information overload in multimedia are one of the prime research topics in the information retrieval (IR) [3,11,13,40]. This is also well illustrated by the existence of numerous competitions, such as TRECVID [4], VBS [19] or LSC [15,44], which focus on a specific sub-domain of multimedia: IR for videos. This has fostered the development of interactive user-in-the-loop systems, exploration of various querying modalities as well as user's feedback models (see, e.g., [1,2,20,28,32,33,36,41,46,48]).

Many of the video retrieval modalities are built on top of the visual similarity of video keyframes. We can mention e.g., query-by-example paradigms [2], incremental relevance feedback models [28], or visual results sorting and clustering [22]. For all the mentioned models, the underlying similarity model's quality is paramount to their final performance. However, most of the tools rely on pre-trained deep learning (DL) models as feature extractors for video keyframes. This brings two possible problems.

First, utilized networks are mostly pre-trained on image domains, such as ImageNet [8], and w.r.t. tasks that substantially diverge from similarity estimation (e.g., image classification [7,10,17,43], or joint text-image models [25,35]). However, the visual appearance of images and video keyframes significantly differs [12]. Most importantly, images often feature a clear focal point (i.e., one main object depicted on a rather simple background), which is not usual for video keyframes. Some evidence for this can be found in Berns et al. [6], who analyzed V3C1 videos and stated that the vast majority of videos contain at least 100 concepts from the 1000 classes of the ImageNet's dataset [39].[1] In addition, video keyframes are often blurry or fuzzy due to the camera or scene motions.

Second, humans may focus on a highly variable set of contextually dependent visual cues [16,24,45] while judging the overall similarity of keyframes. These may substantially differ from the concepts identified by networks pre-trained on, e.g., classification tasks. So, there is a semantic gap between the DL and human understanding of the video keyframes' similarity. The existence of inconsistencies between human-annotated and network-predicted similarity relations was also shown in our preliminary study [47], even for the most up-to-date networks. A natural remedy for this issue is transfer learning [42], but up until now no suitable datasets of human-perceived similarity were available for the video domain.[2]

---

[1] Results of Berns et al. does not state how many of those concepts appear within the same keyframe. Nonetheless, considering the limited volume of keyframes per video in V3C1, we would still argue for higher visual clutteredness of average video keyframes compared to, e.g., ImageNet images.

[2] We previously collected a small dataset to evaluate the quality of pre-trained feature extractors [47], but with only 4K annotated triplets, it is too small for fine-tuning.

Therefore, in this resource paper, we present **RESET** - RElational Similarity Evaluation dataseT, which extends V3C1, a publicly available dataset of general videos [6,38]. RESET contains relative similarity annotations of keyframe triplets (i.e., candidate a is more similar to the query q than candidate b). The main features of the dataset are as follows:

- To cover both the exploration and exploitation phases of video retrieval, RESET contains a mixture of fairly distant as well as highly similar examples w.r.t. multiple similarity metrics. For similar reasons, the dataset features both an unconstrained sample of generic V3C1 videos and two of its compact sub-domains: weddings and diving.
- Most of the triplets were evaluated by multiple users to support the assessments of human consistency. We also collected several contextual features that might affect annotations (e.g., decision time or participant's device).
- For benchmarking purposes, the dataset contains similarity predictions of 30 pre-trained feature extractors.
- The dataset contains in total over 17 thousand similarity annotations. In Sect. 3.3, we show this is already enough to fine-tune some of the best pre-trained models to be more consistent with human-perceived similarity.

1.1 Related Work

Note that the inconsistency of human-perceived and model-predicted similarity is a known problem in visual domains [34,47]. It can be alleviated, e.g., via transfer learning [34,37], or by learning less demanding models from scratch [18]. Nonetheless, to the best of our knowledge, all available datasets focus on images rather than videos [18,23,34,37]. While this may seem like a negligible difference because videos are segmented into keyframes anyway, the appearance characteristics of video keyframes significantly differ from those of utilized image datasets. The existing similarity datasets were collected on top of ImageNet [8], Profiset image collection[3], or THINGS database[4]. In all cases, underlying data contain images that are mostly single-focused on a certain class and generally well-describable (i.e., possess a clear focal point feature). This considerably differs from the typical properties of video keyframes [6,12] and limits the applicability of the image-based similarity datasets for the video domain.

As far as we know, the only exception is our preliminary work [47], where we focused on the consistency of human annotations with pre-trained feature extractors and collected a small dataset (further denoted as RESET-preliminary) of approx. 4K binary annotations. The current work expands over the preliminary one in several directions: a larger dataset containing over 17K similarity annotations was collected, queried triples were repeatedly annotated to ensure consistency, participant device types and decision times were collected, and we also focused on two compact subdomains of V3C1: diving and wedding videos.

---

[3] http://disa.fi.muni.cz/profiset/.
[4] https://things-initiative.org/.

2   RESET Data Collection

In order to receive human annotations, we conducted an extensive web-based user study. We collected users' relative similarity annotations on triplets of a query and two candidate objects (further denoted as $[q, c_1, c_2]$). Figure 1 left contains a screenshot of one of the annotation tasks, where the query image is on top, both candidate items are below, and the user's annotation is expressed via buttons below images.

We utilized the first part of the Vimeo Creative Commons Collection dataset (V3C1) [6,38] as source data, specifically the provided keyframes for each shot, i.e., Master Shot Reference. This dataset was utilized mainly because it is publicly available, contains pre-defined keyframes, and is well-known within the community (among other purposes, it has been utilized in VBS challenge since 2019). We applied a semi-manual cleaning to remove uninformative images (e.g., single-colored screens or very blurry frames), resulting in 1M keyframes in total. We further distinguish three segments of the dataset: keyframes of *wedding* videos, keyframes of scuba *diving* videos, and a *general* set of all keyframes. These were identified w.r.t. presence of corresponding keywords in associated video descriptions or tags. While both *wedding* and *diving* segments are rather small (66373 and 3530 keyframes resp.), they empirically contain a very compact set of visual concepts, and as such, they may be more challenging to judge for both humans and pre-trained models.

2.1   Triplets Generation

While generating $[q, c_1, c_2]$ triplets, we needed to solve the following puzzle: generate examples that are both easy and hard to judge from the human perspective, generate examples where candidates are both close and distant to the query, and do all this without access to human-perceived distance metric upfront. To solve this riddle, we used the distances induced by all of the 30 pre-trained feature extractors to sort the keyframes. As the set of extractors is rather diverse, we believe this approach can serve as a good proxy of human-perceived distance (i.e., at least some of the item pairs considered as similar by the extractors should be perceived as similar by humans too). Then, to reasonably sample both close and distant candidates, we performed an uniform sampling from exponentially increasing bucket sizes. This is motivated by an assumption that only a handful of items are really similar to a given query, so item pairs with smallest distances should be over-sampled as compared to the ones with larger distances.

In particular, consider a fixed query item q, a fixed feature extractor f, and candidate items $c \in C$ representing all keyframes from the *general* data segments. First, candidates are sorted based on their similarity $sim_f(q, c)$ and clustered into one of the buckets. Buckets are exponentially sized with boundaries set to $B = [2^4, 2^8, 2^{12}, 2^{16}, 2^{20}]^5$. Then, we sequentially sample c_1 candidates from all

---

[5] I.e., the third bucket contains items, which are between top-$2^8 + 1$ and top-2^{16} most similar to the query item q.

of the buckets and for each c_1 we sample c_2 candidates from all of the equal-or-more distant buckets. As such, for each q and f, we generate 15 triplets, where some examples have similarly close or distant candidates (i.e., supposedly hard to judge), while in other cases, candidate's distance to the query item highly differs w.r.t. feature extractor f.

The whole procedure is repeated for all feature extractors f and for several randomly selected query items q. The same generation algorithm was also applied for *wedding* and *diving* data segments, but the definition of buckets was slightly altered to align with smaller sets of keyframes. As post-processing, we selected a small volume of *general* triplets as *repeating*, where we aim to collect annotations from more (up to 20) users.

2.2 Utilized Feature Extractors

The utilized feature extractors can be roughly divided into low-level (*color-based* and *SIFT-based*) and high-level (*CNN-based* and *Transformer-based*) classes. The main parameters of all extractors are listed in Fig. 3.

As for **low-level** extractors, we utilized a simple RGB histogram over the pixel representation of the image. We considered each channel separately and used 64 or 256 bins per channel. We also evaluated extractors based on LAB color space [31], namely *LAB Clustered* and *LAB Positional*. *LAB Clustered* clusters all pixels (K-means with $K = 4$) and uses sorted centroids of clusters as image representation. *LAB Positional* creates a regular grid (2×2, 4×4, or 8×8) and computes a mean color for each region. To also enable comparison with low-level beyond-color feature extractors, we utilized VLAD method [21], which essentially post-process the set of scale-invariant feature transform keypoints (SIFT) [30]. For VLAD, we utilized a dictionary size of 64.

For the **high-level** semantic descriptors, we employed several variants of state-of-the-art CNN networks, namely *ResNet* [17], *EfficientNet* [43], *ConvNeXt* [27], *VAN* [14] and *W2VV++* [25]. ResNet, EfficientNet, VAN, and ConvNeXt were trained on the classification task of ImageNet [8] dataset, while W2VV++ utilized MSR-VTT [49] and TGIF [26] datasets to derive joint text-image latent space. In all cases, the penultimate layer was used as the embedding. Furthermore, we also employed several networks based on the transformer architecture, namely Vision transformer (*ViT*) [10], *CLIP* [35] and *ImageGPT* [7]. *ViT* and *ImageGPT* were trained on the ImageNet dataset, while CLIP utilized a large proprietary dataset of 400M image-description pairs. We utilized the penultimate layer for ViT and CLIP, while one of the middle layers was used for ImageGPT (as per the author's suggestions).

2.3 Data Annotation Procedure

The dataset was collected using the annotations of contracted study participants. In contrast to the majority of related approaches, we opted to obtain study participants through the authors' social channels rather than recruiting paid workers. Potential participants were addressed using open calls-for-participation,

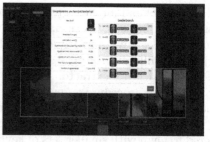

Fig. 1. RESET dataset collection. Left: a screenshot of the data collection step. Right: a screenshot of a "level-up" with user statistics.

which were mostly disseminated to the participants of our previous studies (e.g., [9]), over the friend-of-a-friend axis, and to the students of Charles University.[6]

Each participant was first directed to the study homepage[7], where we asked for some basic demographics and informed consent. Then, a sequence of triplets was presented to the user to annotate which candidate was more similar to the query item. Annotations were expressed on a 5-point Likert scale[8] as illustrated on Fig. 1 left. Note that we did not assume or suggest any particular view of similarity to the participants. The exact judgment prompt was: *"Which image is more similar to the one on the top?"*

Each triplet in the sequence was selected with equal probability from one of *general, wedding, diving,* and *repeated* triplets classes. While for *general, wedding,* and *diving,* the particular triplet is selected at random with equal probability, for *repeated* class, we select the first triplet that has < 20 annotations and was not annotated by the current user yet. In order to account for potential presentation bias issues, the positioning of candidates on the page (i.e., left or right) was randomized.

The volume of annotations per user was bound only by the overall volume of generated triplets, allowing participants to continue the study as long as they wished. Nonetheless, to keep users' attention, the stream was regularly interrupted with "level-ups", where participants' rank, statistics, and the "leaderboard" of most active participants were displayed (see Fig. 1 right).

3   RESET Dataset

The dataset was collected during spring 2023. In total, we recruited 84 participants and obtained 17026 annotations (3719 and 3732 for wedding and diving data segments). Note that the study is ongoing, and we plan to publish extended versions of the dataset in the future. Let us briefly describe the dataset structure and then focus on the data analysis and usage.

---

[6] Participant's demographics are available online from the dataset's homepage.

[7] https://otrok.ms.mff.cuni.cz:8031/user?clang=en.

[8] Exact prompts: *"Left"* - *"Maybe left"* - *"I don't know"* - *"Maybe right"* - *"Right"*.

3.1 Dataset Organization

The RESET dataset is hosted at the Open Science Framework (OSF) and is accessible from https://osf.io/ruh5k/. The dataset (*RESET_data* folder) is comprised of 3 files:

- **participants.csv** store participants' demographics.
- **triplets.csv** store generated triplets: paths to keyframes, details of triplet's generation procedure, and pre-computed distances w.r.t. all extractors.
- **triplet_judgements.csv** store human annotations, their context (time to decision & participant's device), and triplet and participant IDs.

The repository also contains the dataset collected in the preliminary study [47] with an additional 4394 annotated triplets (*"RESET_preliminary"* folder), notebooks covering dataset analysis (*"DataAnalysis"*), source codes to reproduce/extend similarity estimation of all feature extractors (*"FeatureExtractors"*) and a simple fine-tuning framework example (*"ExtractorsFineTuning"*).

3.2 Dataset Analysis

The pool of participants was quite reasonably distributed w.r.t. age, education, and machine learning knowledge.[9] We collected more annotations (78%) from desktop devices, but we still obtained enough annotations from mobile devices to draw some conclusions. For the subsequent analysis, we mapped participants' annotations into $\{-2, -1, 0, 1, 2\}$ values. There were no major differences in the volumes of different annotation values. Specifically, we found no evidence of bias caused by the candidate's position [5].

Participants typically needed only a few seconds to derive the judgment (median: 4.2 s), which is quite understandable given the simplicity of the task. We observed a tendency for participants with more annotations to usually decide faster (Pearson's correlation -0.29), but the difference was not too substantial. We assume this is an artifact of the familiarization process.

Annotations Consistency. We evaluated the consistency among annotators from two perspectives: simplified binary judgments (w.r.t. all pairs of annotations on the same triplet) and a per-triplet standard deviation (STD) of full (5-point Likert scale) annotations. For binary judgments, we observed an overall agreement of 0.80 with some small variations w.r.t. data segments (0.75, 0.82, and 0.81 for wedding, diving, and general, respectively).

For the per-triplet annotations STD, we restricted the dataset to only contain triplets with five or more annotations. The mean per-triplet STD was 0.83 overall, i.e., less than one step in the Likert scale. Again, only minor fluctuation was observed for individual data segments (0.89, 0.79, and 0.79 for wedding, diving, and general segments). Overall, we can conclude that humans have a high degree of agreement in both general and compact data segments.

---

[9] Details were omitted for the sake of space but are available from the repository.

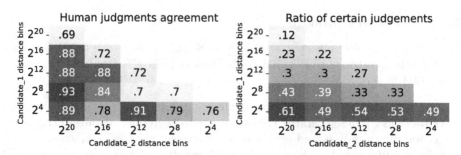

Fig. 2. RESET dataset evaluation. Left: Human binary judgment agreement w.r.t. candidate-query distance bins. Right: ratio of certain annotations w.r.t. candidate-query distance bins.

However, it is fair to note that the judgment was considerably more difficult for more distant candidates. This is illustrated in Fig. 2 (right) for *general* data segment. The figure depicts the ratio of certain annotations (i.e., −2 or 2) as a factor of the distance "bins" to which both candidates belong. Clearly, the more distant the candidates were, the less certain decisions were given. Similarly, the further apart the candidates were from each other, the higher the binary agreement among participants (see Fig. 2 left). Both phenomena were also observed in the wedding and diving segments.[10]

Quality of Feature Extractors. Figure 3 depicts the agreement ratios between predicted annotations and binarized human annotations. The results of the *general* data segment are highly consistent with the preliminary study [47],[11] which we consider as a soft indicator of the study's external validity.

We can observe a rather distinctive drop between the performance of shallow and deep models, with the exception of both ImageGPT variants, whose agreement with human annotations was the lowest of all deep learning models. Seemingly, embeddings based on lower network layers (and thus depicting low-level semantic concepts) are not a suitable representation of human-perceived similarity. Also, we can note that although the best models are not lagging much behind the human agreement levels, there is still space for improvement. While there were some fluctuations in individual data segments, the clear winner in all cases was the *W2VV++* method [25]. Note that W2VV++ differs from all other extractors w.r.t. trained task (except for CLIP, which was also trained to provide joint text-image latent representation) and training data.

Similar results (w.r.t. ordering of extractors) were also obtained while measuring the correlations between full-scale human annotations and the difference in query-candidate distances. Pearson's correlation of the best method (W2VV++) ranged from 0.50 to 0.59 for different data segments.

---

[10] Corresponding figures are available from the RESET dataset repository.

[11] Note that the agreement score in the previous study [47] was slightly lower on average, probably due to the then-employed forced binary decisions.

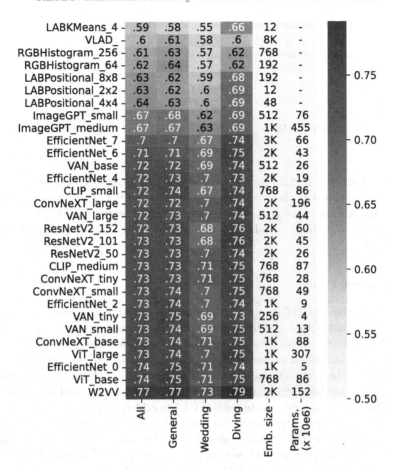

Fig. 3. Agreement ratios between binarized user annotations and predicted annotations by feature extractors. Basic properties of extractors (embeddings size and the volume of trainable parameters) are also depicted.

Context of Human Annotations. In general, almost all extractors were more consistent with the annotations from desktop devices than those from handheld devices (2% drop in average). Only LAB Positional and ImageGPT extractor variants were slightly more consistent with handheld users. We can assume that on smaller displays (thus, with smaller sizes of depicted images), users focus on spanning & more general visual cues rather than on tiny details hidden in the image. This may better correspond to the low-level semantic features. For LAB Positional and ImageGPT, the gain was mostly obtained on the diving data segment, which often features large monochromatic areas suitable, e.g., for grid-based approximations as in LAB Positional.

We also observed some variance in decision times for different annotation values. Highly certain decisions (i.e., -2 and 2) have significantly lower median

Table 1. Fine-tuned cross-validation agreement results

cross-validation set	fine-tuned W2VV++	baseline W2V++
Gen.	μ: **0.788** (σ: 0.019)	μ: 0.771 (σ: 0.026)
Gen. + Div.	μ: **0.799** (σ: 0.008)	μ: 0.785 (σ: 0.010)
Gen. + Wed.	μ: **0.789** (σ: 0.007)	μ: 0.769 (σ: 0.013)
Gen. + Div. + Wed.	μ: **0.797** (σ: 0.010)	μ: 0.781 (σ: 0.009)

time than both less certain and "I don't know" decisions (3.6 vs. 5.0 vs. 4.1 s). The time to decide was also considerably impacted by the device type. Uncertain and less certain decisions (i.e., -1, 0, 1) were faster reached on desktop devices (4.5 vs. 6.1 s), but there was almost no difference on the highly certain decisions (3.6 vs. 3.8 s).

3.3 Dataset Usage

In principle, the RESET dataset has two main use cases: benchmarking additional feature extractors and fine-tuning the existing ones.

In the first case, researchers may expand the list of feature extractors with the ones they are interested in and generate triplet-wise distances w.r.t. these new extractors[12]. Then, *triplets.csv* file should be updated with additional columns corresponding to the newly generated distances, and the data analysis notebooks need to be re-run to incorporate results for the added extractors.

In the latter case, one can utilize the dataset to fine-tune the extractor's weights to better predict human similarity annotations. To exemplify this line of usage, we constructed a simple fine-tuning framework (see *ExtractorsFineTuning folder*) and tested it on the W2VV++ extractor. The fine-tuning process was designed as follows. We aggregated the annotations per triplet and computed the mean from user choices. Fine-tuning was performed with a customized triplet loss

$$\mathcal{L} = max(0, (m + s_f - s_c) * y) + max(0, (|s_f - s_c| - m) * (1 - y))$$

where m is margin and equals to 0.2, y is an absolute value of average user choice, s_c and s_f are predicted similarities between the query item and candidates that should be closer and farther, respectively. The loss function is designed to increase the difference in query-candidate distances when the average user choice is more certain and lower the difference for cases with lower confidence. We used AdamW [29] with weight decay 0.0001, $\beta1 = 0.9$, $\beta2 = 0.999$, and learning rate from cosine decay starting on 0.0001 and with $\alpha = 0$. We employed 5-fold cross-validation and experimented with four different training configurations (data segment combinations). We applied weak data augmentation techniques

---

[12] Please consult the readme in *FeatureExtractors* folder, which describes necessary steps to run the feature extraction pipeline.

(rotations, zooms, and horizontal flips) to increase data variability while still preserving similarity relations.

As can be seen from Table 1, we managed to significantly improve the per-triplet binary agreement[13] with human annotators as compared with the pretrained W2VV++.

4 Conclusions and Future Work

This resource paper describes RESET - RElational Similarity Evaluation dataseT, extending the publicly available V3C1 video collection. The dataset features human similarity judgments on video keyframe triplets. The dataset is primarily suitable for benchmarking and fine-tuning feature extractors to better comply with the human-perceived similarity of videos. As such, it can contribute to the next generation of content-based video retrieval tools based on visual similarity paradigms. The dataset is available from https://osf.io/ruh5k under CC-By Attribution license.

The data collection study is ongoing, and we plan to deploy expanded versions of the dataset regularly. Apart from continued data collection, we plan to focus primarily on the fine-tuning applications of the dataset. Some early experiments confirmed that the current RESET dataset is already large enough to secure some improvements. However, in the future, we plan to focus also on ensembles of multiple models, explore different ways to utilize multi-valued feedback in the learning process, and focus on enhancing the variability of the data, e.g., by also employing RESET_preliminary or image-based datasets (e.g., in the early fine-tuning stages). In another direction of our future work, we would like to expand on the device context and check to what extent users' concept of similarity differs when multiple (small) images are involved.

Acknowledgement. This paper has been supported by Czech Science Foundation (GAČR) project 22-21696S. Computational resources were provided by the e-INFRA CZ project (ID:90254),supported by the Ministry of Education, Youth and Sports of the Czech Republic.

References

1. Alam, N., Graham, Y., Gurrin, C.: Memento 2.0: an improved lifelog search engine for LSC'22. In: Proceedings of the 5th Annual on Lifelog Search Challenge, LSC '22, pp. 2–7. ACM (2022). https://doi.org/10.1145/3512729.3533006
2. Andreadis, S., et al.: VERGE in VBS 2022. In: Por Jonsson, B., et al. (eds.) MultiMedia Modeling. Lecture Notes in Computer Science, vol. 13142, pp. 778–783. Springer, Cham (2020). https://doi.org/10.1007/978-3-030-98355-0_50
3. Asim, M.N., Wasim, M., Ghani Khan, M.U., Mahmood, N., Mahmood, W.: The use of ontology in retrieval: a study on textual, multilingual, and multimedia retrieval. IEEE Access **7**, 21662–21686 (2019). https://doi.org/10.1109/ACCESS.2019.2897849

---

[13] All per-triplet human judgments were averaged and then binarized.

4. Awad, G., et al.: An overview on the evaluated video retrieval tasks at TRECVID 2022. arXiv (2023)
5. Bar-Ilan, J., Keenoy, K., Levene, M., Yaari, E.: Presentation bias is significant in determining user preference for search results-a user study. J. Am. Soc. Inform. Sci. Technol. **60**(1), 135–149 (2009). https://doi.org/10.1002/asi.20941
6. Berns, F., Rossetto, L., Schoeffmann, K., Beecks, C., Awad, G.: V3c1 dataset: an evaluation of content characteristics. In: Proceedings of the 2019 on International Conference on Multimedia Retrieval, ICMR '19, pp. 334–338. ACM (2019). https://doi.org/10.1145/3323873.3325051
7. Chen, M., et al.: Generative pretraining from pixels. In: ICML'20. PMLR (2020)
8. Deng, J., Dong, W., Socher, R., Li, L.J., Li, K., Fei-Fei, L.: ImageNet: a large-scale hierarchical image database. In: CVPR'09, pp. 248–255. IEEE (2009)
9. Dokoupil, P., Peska, L.: LiGAN: recommending artificial fillers for police photo lineups. In: 3rd Knowledge-aware and Conversational Recommender Systems. KaRS@RecSys 2021, vol. 2960. CEUR-WS.org (2021). http://ceur-ws.org/Vol-2960/paper14.pdf
10. Dosovitskiy, A., et al.: An image is worth 16x16 words: transformers for image recognition at scale. arXiv (2020)
11. Gasser, R., Rossetto, L., Schuldt, H.: Towards an all-purpose content-based multimedia information retrieval system. arXiv (2019)
12. Gauen, K., et al.: Comparison of visual datasets for machine learning. In: 2017 IEEE International Conference on Information Reuse and Integration (IRI), pp. 346–355 (2017). https://doi.org/10.1109/IRI.2017.59
13. Georgiou, T., Liu, Y., Chen, W., Lew, M.: A survey of traditional and deep learning-based feature descriptors for high dimensional data in computer vision. Int. J. Multimedia Inf. Retrieval **9**(3), 135–170 (2020). https://doi.org/10.1007/s13735-019-00183-w
14. Guo, M.H., Lu, C.Z., Liu, Z.N., Cheng, M.M., Hu, S.M.: Visual attention network. Comput. Vis. Media **9**(4), 733–752 (2023). https://doi.org/10.1007/s41095-023-0364-2
15. Gurrin, C., et al.: Comparing approaches to interactive lifelog search at the lifelog search challenge (LSC2018). ITE Trans. Media Technol. Appl. **7**(2), 46–59 (2019)
16. Han, S., Humphreys, G.W., Chen, L.: Uniform connectedness and classical gestalt principles of perceptual grouping. Percept. Psychophys. **61**(4), 661–674 (1999). https://doi.org/10.3758/BF03205537
17. He, K., Zhang, X., Ren, S., Sun, J.: Identity mappings in deep residual networks. In: Leibe, B., Matas, J., Sebe, N., Welling, M. (eds.) Computer Vision - ECCV 2016. Lecture Notes in Computer Science(), vol. 9908, pp. 630–645. Springer, Cham (2016)
18. Hebart, M.N., Zheng, C.Y., Pereira, F., Baker, C.I.: Revealing the multidimensional mental representations of natural objects underlying human similarity judgements. Nat. Hum. Behav. **4**(11), 1173–1185 (2020)
19. Heller, S., et al.: Interactive video retrieval evaluation at a distance: comparing sixteen interactive video search systems in a remote setting at the 10th video browser showdown. Int. J. Multimed. Inf. Retr. **11**(1), 1–18 (2022). https://doi.org/10.1007/s13735-021-00225-2
20. Hezel, N., Schall, K., Jung, K., Barthel, K.U.: Efficient search and browsing of large-scale video collections with vibro. In: Por Jonsson, B., et al. (eds.) MultiMedia Modeling. Lecture Notes in Computer Science, vol. 13142, pp. 487–492. Springer, Cham (2022). https://doi.org/10.1007/978-3-030-98355-0_43

21. Jégou, H., Douze, M., Schmid, C., Pérez, P.: Aggregating local descriptors into a compact image representation. In: CVPR'10, pp. 3304–3311. IEEE (2010)
22. Jung, K., Barthel, K.U., Hezel, N., Schall, K.: PicArrange - visually sort, search, and explore private images on a mac computer. In: Por Jonsson, B., et al. (eds.) MultiMedia Modeling. Lecture Notes in Computer Science, vol. 13142, pp. 452–457. Springer, Cham (2022). https://doi.org/10.1007/978-3-030-98355-0_38
23. Křenková, M., Mic, V., Zezula, P.: Similarity search with the distance density model. In: Skopal, T., Falchi, F., Lokoc, J., Sapino, M.L., Bartolini, I., Patella, M. (eds.) Similarity Search and Applications. Lecture Notes in Computer Science, vol. 13590, pp. 118–132. Springer, Cham (2022). https://doi.org/10.1007/978-3-031-17849-8_10
24. Kubovy, M., Berg, M.: The whole is equal to the sum of its parts: a probabilistic model of grouping by proximity and similarity in regular patterns. Psychol. Rev. **115**, 131–54 (2008). https://doi.org/10.1037/0033-295X.115.1.131
25. Li, X., Xu, C., Yang, G., Chen, Z., Dong, J.: W2VV++ fully deep learning for Ad-hoc video search. In: ACM MM'19, pp. 1786–1794 (2019)
26. Li, Y., Song, Y., Cao, L., Tetreault, J., Goldberg, L., Jaimes, A., Luo, J.: TGIF: a new dataset and benchmark on animated GIF description. In: CVPR'16, pp. 4641–4650 (2016)
27. Liu, Z., Mao, H., Wu, C.Y., Feichtenhofer, C., Darrell, T., Xie, S.: A convnet for the 2020s. In: Proceedings of the IEEE/CVF Conference on Computer Vision and Pattern Recognition, pp. 11976–11986 (2022)
28. Lokoč, J., Mejzlík, F., Souček, T., Dokoupil, P., Peška, L.: Video search with context-aware ranker and relevance feedback. In: Por Jonsson, B., et al. (eds.) MultiMedia Modeling. Lecture Notes in Computer Science, vol. 13142, pp. 505–510. Springer, Cham (2022). https://doi.org/10.1007/978-3-030-98355-0_46
29. Loshchilov, I., Hutter, F.: Decoupled weight decay regularization. arXiv (2019)
30. Lowe, D.G.: Distinctive image features from scale-invariant keypoints. Int. J. Comput. Vision **60**(2), 91–110 (2004)
31. McLaren, K.: The development of the CIE 1976 (L*A*B*) uniform colour-space and colour-difference formula. J. Soc. Dye. Colour. **92**, 338–341 (2008)
32. Nguyen, T.N., et al.: LifeSeeker 4.0: an interactive lifelog search engine for LSC'22. In: Proceedings of the 5th Annual on Lifelog Search Challenge, LSC '22, pp. 14–19, pp. 14–19. ACM (2022). https://doi.org/10.1145/3512729.3533014
33. Peška, L., Kovalčík, G., Souček, T., Škrhák, V., Lokoč, J.: W2VV++ BERT model at VBS 2021. In: Lokoc, J., et al. (eds.) MultiMedia Modeling. Lecture Notes in Computer Science(), vol. 12573, pp. 467–472. Springer, Cham (2021). https://doi.org/10.1007/978-3-030-67835-7_46
34. Peterson, J.C., Abbott, J.T., Griffiths, T.L.: Evaluating (and improving) the correspondence between deep neural networks and human representations. Cogn. Sci. **42**(8), 2648–2669 (2018)
35. Radford, A., et al.: Learning transferable visual models from natural language supervision. In: ICML'19, pp. 8748–8763. PMLR (2021)
36. Ribiero, R., Trifan, A., Neves, A.J.R.: MEMORIA: a memory enhancement and moment retrieval application for LSC 2022. In: Proceedings of the 5th Annual on Lifelog Search Challenge, LSC '22, pp. 8–13. ACM (2022). https://doi.org/10.1145/3512729.3533011
37. Roads, B.D., Love, B.C.: Enriching ImageNet with human similarity judgments and psychological embeddings. In: CVPR'21, pp. 3547–3557. IEEE/CVF (2021)

38. Rossetto, L., Schuldt, H., Awad, G., Butt, A.A.: V3C - a research video collection. In: Kompatsiaris, I., Huet, B., Mezaris, V., Gurrin, C., Cheng, W.H., Vrochidis, S. (eds.) MultiMedia Modeling. Lecture Notes in Computer Science(), vol. 11295, pp. 349–360. Springer, Cham (2019). https://doi.org/10.1007/978-3-030-05710-7_29

39. Russakovsky, O., et al.: ImageNet large scale visual recognition challenge. Int. J. Comput. Vision (IJCV) **115**(3), 211–252 (2015). https://doi.org/10.1007/s11263-015-0816-y

40. Saxena, P., Singh, S.K., Srivastava, M.: Content-based retrieval of multimedia information using multiple similarity indexes. In: Pant, M., Sharma, T., Verma, O., Singla, R., Sikander, A. (eds.) Soft Computing: Theories and Applications. Advances in Intelligent Systems and Computing, vol. 1053, pp. 1235–1242. Springer, Singapore (2020). https://doi.org/10.1007/978-981-15-0751-9_113

41. Sungjune, P., Song, J., Park, M., Ro, Y.M.: IVIST: interactive video search tool in VBS 2020. In: Ro, Y., et al. (eds.) MultiMedia Modeling. Lecture Notes in Computer Science(), vol. 11962, pp. 809–814. Springer, Cham (2020). https://doi.org/10.1007/978-3-030-37734-2_74

42. Tan, C., Sun, F., Kong, T., Zhang, W., Yang, C., Liu, C.: A survey on deep transfer learning. In: Kurkova, V., Manolopoulos, Y., Hammer, B., Iliadis, L., Maglogiannis, I. (eds.) Artificial Neural Networks and Machine Learning - ICANN 2018. LNCS(), vol. 11141, pp. 270–279. Springer, Cham (2018). https://doi.org/10.1007/978-3-030-01424-7_27

43. Tan, M., Le, Q.: EfficientNet: rethinking model scaling for convolutional neural networks. In: ICML'19, pp. 6105–6114. PMLR (2019)

44. Tran, L.D., et al.: Comparing interactive retrieval approaches at the lifelog search challenge 2021. IEEE Access (2023)

45. Triesch, J., Ballard, D.H., Jacobs, R.A.: Fast temporal dynamics of visual cue integration. Perception **31**(4), 421–434 (2002). https://doi.org/10.1068/p3314. PMID: 12018788

46. Veselý, P., Mejzlík, F., Lokoč, J.: SOMHunter V2 at video browser showdown 2021. In: Lokoc, J., et al. (eds.) MultiMedia Modeling. LNCS(), vol. 12573, pp. 461–466. Springer, Cham (2021). https://doi.org/10.1007/978-3-030-67835-7_45

47. Veselý, P., Peška, L.: Less is more: similarity models for content-based video retrieval. In: Dang-Nguyen, D.T., et al. (eds.) MultiMedia Modeling. LNCS, vol. 13834, pp. 54–65. Springer, Cham (2023). https://doi.org/10.1007/978-3-031-27818-1_5

48. Wu, J., Nguyen, P.A., Ma, Z., Ngo, C.W.: SQL-Like interpretable interactive video search. In: MultiMedia Modeling, pp. 391–397 (2021)

49. Xu, J., Mei, T., Yao, T., Rui, Y.: MSR-VTT: a large video description dataset for bridging video and language. In: CVPR'16, pp. 5288–5296 (2016)

A New Benchmark and OCR-Free Method for Document Image Topic Classification

Zhen Wang[1(✉)], Peide Zhu[2], Fuyang Yu[3], and Manabu Okumura[1]

[1] Tokyo Institute of Technology, Tokyo, Japan
wzh@lr.pi.titech.ac.jp, oku@pi.titech.ac.jp
[2] Delft University of Technology, Delft, Netherlands
p.zhu-1@tudelft.nl
[3] Beihang University, Beijing, China
yfy1996@buaa.edu.cn

Abstract. Document image understanding is challenging, given the complexity of the combination of illustrations and text that make up a document image. Previous document image classification datasets and models focus more on the document format while ignoring the meaningful content. In this paper, we introduce DocCT, the first-of-its-kind content-aware document image classification dataset that covers various daily topics that require understanding fine-grained document content to perform correct classification. Furthermore, previous pure vision models cannot sufficiently understand the semantic content of document images. Thus OCR is commonly adopted as an auxiliary component for facilitating content understanding. To investigate the possibility of understanding document image content without the help of OCR, we present DocMAE, a new self-supervised pretrained document image model without any extra OCR information assistance. Experiments show that DocMAE's ability to understand fine-grained content is far greater than previous vision models and even surpasses some OCR-based models, which proves that it is possible to well understand the semantics of document images only with the help of pixels. (Dataset can be downloaded at https://github.com/zhenwangrs/DocCT).

Keywords: Document Image · Pretraining · Topic Classification

1 Introduction

The task of visual document understanding (VDU) aims at automatically reading and understanding document images. Digital images of documents are an important source of information; for example, in digital libraries, documents are often stored as scanned images before further processing such as optical character recognition (OCR) [9]. A document image contains rich content elements like text, images, and diagrams, organized in various styles, which makes VDU a challenging task. One important task toward visual document understanding is

---

Z. Wang and P. Zhu—Equal Contribution.

S. Rudinac et al. (Eds.): MMM 2024, LNCS 14565, pp. 15–27, 2024.
https://doi.org/10.1007/978-3-031-56435-2_2

document image classification (DIC), which aims to classify a document image into a category, similar to vanilla image classification like ImageNet [6], or multimodal classification dataset like N24News [20]. DIC can be used in various applications, such as automatic book classification in the library, helping Internet search engines better integrate different information, or determining which domain-specific model should be used for OCR. It is also an essential step toward a more fine-grained understanding of document images, which can inspire some downstream tasks such as document visual question answering [16] (Fig. 1).

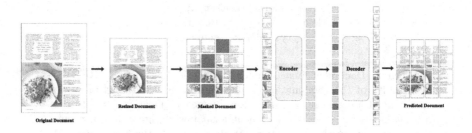

Fig. 1. The overall pipeline of DocMAE consists of an encoder and a decoder, mainly following the architecture of MAE [10]. The input document image is first resized to 640 × 640 and then split into numbers of patches. Some patches are masked by a certain ratio. Then the unmasked patches are concatenated to a sequence and fed into the transformer encoder. The masked patches and the output of the encoder are combined together and sent to the transformer decoder to predict the pixel of the masked patches.

Most previous DIC research relies on RVL-CDIP [9], a large-scale dataset that categorizes document images into 16 classes like "email", "invoice", and "magazine", based on their formats. However, it pays little attention to the document image's concrete content, while semantics conveyed by the content is also essential. For example, we would rather know what topic the document image talks about than whether it is an email. Further, the data in RVL-CDIP are all on a similar topic, which makes them indistinguishable for fine-grained content topic classification. The lack of suitable datasets has become an obstacle to further developing DIC methods for content type classification.

Therefore, in this paper, to facilitate further DIC research, we present the first document image dataset that contains fine-grained topic annotations: **DocCT** (***Doc**ument Image **C**lassification via **T**opic*). In DocCT, there are 10 categories, all of which are common topics in daily life. Each category contains documents in various formats. DocCT can prompt models' document image content understanding ability since the model can only classify the document image correctly when it can capture the correct features of the image's content.

With DocCT, we then evaluate some state-of-the-art models developed for document image classification. Current DIC methods can be summarized into two categories. One category contains pure vision methods that adopt models like CNN [9] or transformers [12,15] for image classification, usually used in document format or layout analysis. The other category contains methods that adopt a two-stream multimodal framework. These methods first extract text in the document image by OCR and then perform classification with both OCR-text and image features [2,11]. The performance of methods in this category is heavily restricted by the quality of text extracted by OCR. Our experiments on DocCT reveal a huge performance gap between DIC methods in both categories from humans, proving that document image understanding is still challenging. DocCT is thus a good benchmark for model evaluation.

To mitigate this gap, in this paper, we further investigate the possibility of developing an effective method for the content-based document image classification with the pure vision method. To this end, we present a novel self-supervised pretrained model - **DocMAE** (**Doc**ument **M**asked **A**uto**E**ncoder), which is trained with large-scale unlabeled document images. In DocMAE, we enlarge the input image size to better understand the semantics of text composed of pixels. Experimental results on DocCT demonstrate that this adjustment dramatically improves the model's ability to recognize a fine-grained semantic topic in images, thus significantly surpassing previous models, even OCR-based methods, in classification, making it more suitable for content-aware tasks.

Our contributions can be summarized as follows: (1) We present DocCT, the first DIC dataset with fine-grained content type annotations for document image topic classification tasks. (2) We present DocMAE, the first self-supervised pretrained model with a deeper understanding of image content. (3) Our experimental results reveal some unique challenges from DocCT. Further, with DocMAE, we prove that the model can understand the document image content by pixels without explicitly extracting its text by OCR no matter in pretraining or fine-tuning.

2   Related Work

2.1   Document Image Classification

With the development of deep image models, document image related research is attracting more attention. Compared to vanilla image research on ImageNet [6], document images are more complex, given their much richer content. As an important task for the document images, DIC [4] is one of the earliest and most researched directions. In DIC, a given document image should be classified into a correct category by specific requirements. The most widely used dataset is RVL-CDIP [9], a sub-dataset of IIT-CDIP [14]. The images in IIT-CDIP are scanned documents collected from the public records of lawsuits against American tobacco companies. RVL-CDIP contains 16 document formats, such as "letter" or "invoice", which can be used to evaluate models' classification ability.

However, compared to recognizing the format of an image, understanding its content is more critical and challenging, since it can facilitate lots of higher-level AI research such as visual question answering [1]. Thus in this paper, we present

DocCT, the first DIC dataset that focuses on document content understanding, hoping to prompt research in related fields.

2.2 Pretraining Document Models

The goal of pretraining technologies is to use a large amount of unsupervised text to pretrain a model, so that the model can master prior knowledge, improving the performance of downstream tasks. After the success of ViT [8], which first applies vanilla transformer [18] to vision tasks, researchers start to investigate how to better pretrain ViT in image-related tasks like BERT [7] in natural language processing. Currently, there are also some pretrained models for document image related research. DiT [15] is a pretrained document model based on BEiT [3]. Donut [12] is an encoder-decoder model which uses the vanilla transformer-based architecture. It does not use OCR while fine-tuning, but it uses OCR in the supervised pretraining phase to let the model know how to output the correct text from the input image.

Some document models convert document image tasks into a multimodal task, such as LayoutLM [11], DocFormer [2], and LiLT [19]. They use OCR to extract the text information from a document image and input both the original image and OCR text into the models. Compared to pure image models, they can obtain higher accuracy with the extra text input, while the training process is time-consuming and inefficient in making an inference.

However, most previous pretrained document models aim at document layout analysis, making them unsuitable for solving fine-grained document content understanding when applied to datasets like DocCT. Thus in this paper, we present DocMAE, a large-scale self-supervised pretrained model. It is a pure image model like DiT without OCR, while it is also helpful in understanding the semantic information in the image and can be further used in other document-related downstream tasks.

3 DocCT Dataset

3.1 Data Collection

We collected our dataset from web images with search engines. To cover as many topics as possible, we started from the root node of the wiki's category tree and selected the 10 most commonly seen topics in our daily life, including "Artist", "Buildings", "Economy", "Education", "Food", "Entertainment", "Environment", "Sports", "Health", and "Technology".

For each category, to ensure that most search engine search results are relevant documents, we constructed our search keywords with the category name alongside diverse document format names. As for the document format, we first adopted 16 types in RVL-DCIP and then added some novel formats to cover as many formats as possible. Finally, we settled on a total of 27 types of formats, including "book", "budget", "contract", "email", "exam", "flow chat", "form", "introduction", "invoice", "letter", "magazine", "map", "memo", "newspaper", "phone

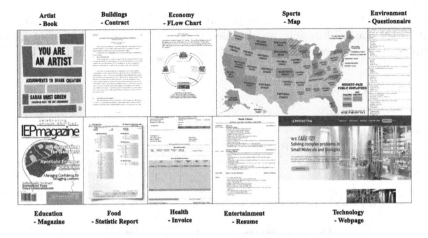

| Artist - Book | Buildings - Contract | Economy - FLow Chart | Sports - Map | Environment - Questionnaire |

| Education - Magazine | Food - Statistic Report | Health - Invoice | Entertainment - Resume | Technology - Webpage |

Fig. 2. 10 categories and some of the formats in DocCT.

application", "poster", "presentation slides", "print advertisement", "questionnaire", "resume", "scientific publication", "specification", "statistical repost", "textbook", and "webpage". For each format, we collected up to 300 images. With those topics and formats as the search keywords, we crawled nearly 80K images in the collection procedure (Fig. 2).

3.2 Annotation and Quality Control

Table 1. Statistics of DocCT. The total number of document images is 23512.

Category	#Count	Category	#Count
Artist	2531	Buildings	2089
Economy	2603	Education	2609
Food	3301	Entertainment	1984
Sports	2541	Environment	1544
Health	2032	Technology	2278

We then asked crowdworkers to annotate the crawled images. Given an image, the annotating procedure is as follows:

- **Step 1**: Determine whether the image is a document image. An image without any text information or with too vague text to recognize will be dropped.
- **Step 2**: Determine whether the document image conforms to the corresponding category. The irrelevant image will be removed. If an image can belong to more than one category, it will also be discarded.

Only images that pass the above judgments will be considered valid and be kept. After manual filtering, we obtained about 23K accurate document image samples. In Table 1, we provide the statistics for each category in DocCT.

3.3   Data Analysis

| Artist - | Buildings - | Food - | Sports - |
| Resume | Resume | Questionnaire | Questionnaire |

Fig. 3. Comparison between two categories with the same format.

With lots of different formats, DocCT is able to reflect common knowledge content that documents in different formats can narrate under the same topic in our daily life, making the research on it more applicable. Compared with RVL-CDIP, the formats we chose contain more modern and diverse document formats with more vivid colors than a single white-black scanned file. In Fig. 3, we present comparisons between two different categories with the same format. In DocCT, the layout of two categories with the same format is very similar. This can ensure that models cannot cheat with layouts and must analyze the detailed content. Models can yield correct classification only through understanding the semantics conveyed by a document image.

4   DocMAE

4.1   Architecture

Unlike DiT and LayoutLM, which use BEiT [3] as the visual backbone, in this paper, we choose MAE [10] as the basic architecture of DocMAE. Compared with BEiT, using dVAE [17] to tokenize image patches first, MAE directly uses pixel reconstruction to calculate model prediction loss. This is a better choice for a document image since the pixels of a document image are more complex and contain more semantics. It is difficult to represent all cases with a limited number of tokens (8192 tokens used in BEiT).

[5] proves that as an important part of MAE, the decoder can steal some abilities from the encoder, which will significantly limit the encoder's ability when only using the encoder to do a downstream task. Thus, unlike the original MAE, DocMAE keeps both the encoder and decoder when fine-tuning to ensure better performance.

4.2   Pretraining Settings

We used MAE$_{base}$ as the basic architecture of DocMAE. The DocMAE encoder is a 12-layer transformer with 768 hidden size and 12 attention heads. The feed-forward network size is 3072. The DocMAE decoder is a 7-layer transformer with a hidden size 512 and 16 attention heads. The feed-forward network size is 2048. The input image size is 640 × 640, and we employed 20 × 20 as the patch size. A special [CLS] token was concatenated to the start of the patch sequence. The mask ratio was set to 30%, which means that in pretraining, while the input sequence length to the encoder is 718 ($717+1$), the input sequence length to the decoder is 1025 ($1024+1$). To make DocMAE adapt to documents of different original resolutions and shapes, we randomly cropped the input images with 10% probability during pretraining.

We choose 30% as the mask ratio because a document image, which contains plenty of paragraphs and various illustrations, is much more complex than an object image such as a cat or dog. The 75% mask ratio of the original MAE is too high and will eliminate lots of useful information for the model to predict. Thus we chose 30%, a value between 75% (MAE) and 15% (BERT [7]).

We pretrained DocMAE for 100 epochs with 512 batch size. The optimizer is Adam [13] with $\beta_1 = 0.9$ and $\beta_2 = 0.999$, and weight decay is 0.05. The start learning rate is 1e-4 with cosine annealing learning rate decay and without warmup. The dropout was disabled. The whole pretraining procedure lasted three weeks with four RTX 3090 GPUs.

4.3   Pretraining Corpus

To make DocMAE applicable to more diverse tasks, unlike DiT and LayoutLM, which directly use documents from IIT-CDIP, we used open-domain magazines as the pretraining corpus since magazines contain various document types, including both plain and rich text. We collected massive magazines and converted each magazine into a collection of document images. In total, we collected around 1.6 million open-domain document images. Since the collection method of the pretraining corpus is different from DocCT, we added an additional data filter to remove the data duplication between them.

4.4   Input Size Setting

Input size plays an essential role in a deep learning image model since too small image size will lead to loss of information, while too large image size will make it difficult to train the model. To balance training time and information retention, almost all previous image models chose 224 × 224 as the input image size. This image size has achieved excellent results on object image classification datasets such as ImageNet [6]. Thus, some document image pretrained models, such as DiT and LayoutLM, also chose this size for the input.

However, our investigation showed that 224 is not an appropriate size for images with text information such as document images. The small input size will

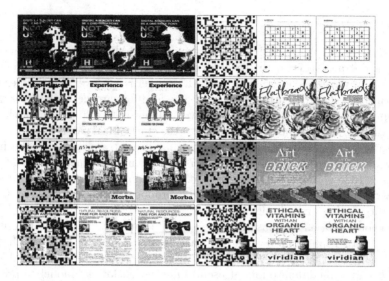

Fig. 4. Image restoration for some document images. From left to right are the **masked image**, **restored image**, and **original image**. The mask ratio is 30%.

lead to the loss of text information. This may have small effects on identifying whether an animal is a cat or dog or on determining the layout of a table in a document. However, if we want the model to identify more fine-grained text semantics in an image, the expansion of the input size is required since a word is much smaller in an image than a cat or table. We chose 640 as the input image size, ensuring that the text in most document images is recognizable while still applicable for training the model. The comparison of 224×224 and 640×640 is shown in Fig. 5. The image with 640 size proves to contain richer and clearer text information either in plain or rich text images.

4.5   Evaluation

After pretraining, we used DocMAE to restore some document images randomly searched on the Internet. Some of the results are shown in Fig. 4. We inputted a picture with 30% of the random area masked and observed the output. It can be found that the overall image can be restored relatively well, and the restoration for larger texts is excellent. However, for texts with small font size, the restoration is still blurred. This shows that DocMAE still has some room for improvement.

5   Experiments on DocCT

5.1   Experimental Settings

DocCT was split into training, validation, and test sets with a ratio of 8:1:1. We used the training set to train the model, took the best model on the validation

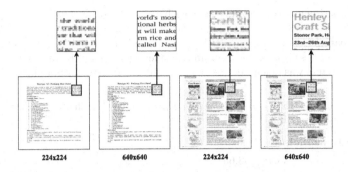

Fig. 5. Comparison between 224 × 224 and 640 × 640. The left is a plain text image while the right is a rich text image. In either case, the image with the size of 224 loses most of the text information, while the image with 640 keeps the text legible.

set, and then recorded its performance on the test set. We evaluated DocMAE in three ways. One is DocMAE$_{encoder}$, which uses only the encoder of DocMAE. The other is DocMAE$_{decoder}$, which fixes the parameters of the encoder and fine-tunes only the decoder. The last DocMAE$_{full}$ is to fine-tune all the parameters in the encoder and decoder. Compared models are mainly divided into two categories. One is image-only models that depend entirely on the processing of pixels, including BEiT [3], DiT [15], and MAE [10]. The other is OCR-enhanced multimodal models with text extracted by OCR as the additional input; here, we chose LayoutLMv3 [11].

5.2   Classification Accuracy

In the document content classification task on DocCT, DocMAE achieves the best performance among all image-only models. DocMAE$_{full}$ obtains a comparable result to OCR-based methods such as LayoutLMv3 (74.54 vs. 76.94 in F1), and DocMAE$_{decoder}$ greatly excels at OCR-based methods, demonstrating that it is possible to capture semantic information by using only pixel data in document images instead of directly using OCR.

It can also be found that MAE obtains a much higher F1 than DiT, proving that direct pixel prediction as a pretraining task is better for understanding document semantics than token prediction used in BEiT and DiT. This is mainly because text pixels are more complex, and it is difficult to summarize all the possible image patches by just using 8192 tokens.

Compared with Donut and DocMAE, though both are pure vision models, Donut is a generation model that uses OCR to extract the text while doing supervised pretraining. The higher accuracy of DocMAE proved that it is possible for the vision model to achieve considerable performance only with OCR-free unsupervised pretraining.

Furthermore, we randomly selected 500 images for human annotators to classify, and the accuracy of human beings is 98.60%, which is much higher than the current deep learning models. It shows that the models still have a lot of room

Table 2. Experimental results on DocCT with different models. DocMAE$_{encoder}$ means we utilized only the DocMAE encoder for the classification model. DocMAE$_{decoder}$ means the parameters in the encoder were fixed, and only the decoder was fine-tuned. DocMAE$_{full}$ means both the encoder and decoder were used to be fine-tuned. Training and inference time was calculated on a single RTX3090 GPU within one epoch.

Model	F1	ACC	Train/Epoch	Infer/Epoch	Image Size	#Param
Human	–	98.60	–	–	–	–
Image-Only Models						
BEiT$_{base}$	38.48	38.65	2 m 32 s	30 s	224	87 M
DiT$_{base}$	39.89	39.92	2 m 30 s	31 s	224	87 M
DiT$_{large}$	43.58	43.95	4 m 47 s	40 s	224	304 M
MAE$_{base(encoder)}$	41.92	42.00	2 m 31 s	31 s	224	87 M
MAE$_{base(decoder)}$	41.22	40.94	2 m 20 s	30 s	224	113 M
MAE$_{base(full)}$	42.17	42.68	3 m 05 s	35 s	224	113 M
DocMAE$_{224(encoder)}$	45.10	45.86	2 m 31 s	30 s	224	87 M
DocMAE$_{224(full)}$	46.76	47.09	3 m 05 s	35 s	224	113 M
DocMAE$_{224(decoder)}$	46.94	47.60	2 m 20 s	30 s	224	113 M
Donut	82.27	82.13	39 m 18 s	14 m 57 s	640	202 M
DocMAE$_{encoder}$	36.30	37.46	17 m 40 s	1 m 01 s	640	87 M
DocMAE$_{full}$	74.54	74.55	31 m 13 s	1 m 19 s	640	113 M
DocMAE$_{decoder}$	**83.53**	**83.84**	14 m 22 s	1 m 17 s	640	113 M
OCR-Enhanced Models						
LayoutLMv3$_{base}$	75.63	75.64	51 m 51 s	7 m 05 s	224	133 M
LayoutLMv3$_{large}$	76.94	76.91	55 m 32 s	7 m 11 s	224	368 M

for improvement, and DocCT proves to be a challenging dataset that is worth researching.

5.3   Encoder vs. Decoder

We then performed ablation analysis for different parts of the DocMAE architecture to observe the effect of the different modules on the accuracy. We first fine-tuned DocMAE only with the encoder. Compared to full DocMAE, the DocMAE encoder obtains only 36.30 in F1. This vast performance drop proves that in dealing with document images, the decoder is an essential part and cannot be removed as MAE does for ImageNet.

Another interesting finding is that, when the encoder module of DocMAE is fixed and only the decoder module is fine-tuned, the model obtains even higher accuracy (83.53 vs. 74.54 in F1). We think this phenomenon is because, when DocMAE is fully pretrained, the encoder can already extract the features of a document image well. Any further fine-tuning of the encoder will affect the feature extraction ability, thus affecting the overall accuracy. Document images are more complex than images of simple objects, making this disturbance more obvious. Our experiments prove that in DocMAE, the encoder is suitable for acting as a feature extractor, while the decoder can be used for migrating to downstream tasks.

5.4   Influence of Resolution

To confirm that the input image resolution does affect the model's understanding of the semantics in a document image, we additionally pretrained a model named $DocMAE_{224}$ with the same settings as DocMAE. The only difference is that the input image size of $DocMAE_{224}$ is 224×224. The experimental comparison results are shown in Table 2. Although the performance of $DocMAE_{224}$ on DocCT is much better than the original MAE with the help of pretraining based on document image data, there is still a huge gap compared to DocMAE with 640 image resolution (46.94 vs. 83.53 in F1). This result effectively proves that larger resolution is crucial for the semantic understanding of document images.

5.5   Model Efficiency

Since the model structure of different methods varies, we also recorded the efficiency of the different models during training and inference. Compared with DiT_{base}, $DocMAE_{full}$ is much slower (31m13s vs. 2m30s), because, as the length of input image patches increases (1025 vs. 197), the training time also increases exponentially. However, when it comes to inference, DocMAE is not much slower than DiT_{base} (1m19s vs. 31 s) and DiT_{large} (1m19s vs. 40 s). And since Donut is designed to be a generative model, it is much slower both at training and inference. As for the OCR-based methods, they are the slowest among all methods, both during training and inference. $DocMAE_{full}$ takes half as long to train an epoch as LayoutLMv3 and reaches even a speed of nearly 6 times in inference. This is mainly because OCR is time-consuming no matter in training or inference. DocMAE is proved to be a practical model that is well suited for solving document image related tasks by comparing all methods, including both the OCR-free and OCR-based methods. It has better accuracy than DiT while it also has higher efficiency than the OCR-based methods.

5.6   Error Analysis

To gain an intuitive perception of the features of cases where the model works or where it does not, we performed error analysis for several cases. We chose DiT_{large}, $LayoutLMv3_{large}$, and $DocMAE_{decoder}$ to compare, and their results are shown in Fig. 6.

   In the first case, all three models can classify it correctly. There are apparent objects and keywords in the image. Since the compression of the input image resolution will not lose important information, even the OCR does successfully extract the correct text. In the second case, when there is no significant object and full of fine-grained text, due to the small image size, DiT is not able to

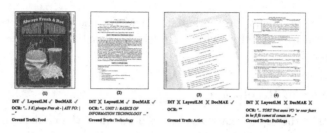

Fig. 6. Classification results on the test set with DiT$_{large}$, LayoutLMv3$_{large}$, and DocMAE$_{decoder}$. ✓ indicates correct classification and × indicates incorrect classification. The OCR results come from LayoutLMv3.

recognize deep semantic information and just fails. However, in spite of the same image size, since LayoutLMv3 has OCR as a complement input, it can obtain enough meaningful information directly from the OCR text and thus can still classify it correctly. In the third case, because the text is relatively small and skewed, OCR cannot precisely extract the text, making the final classification result of LayoutLMv3 wrong. Those cases prove that DocMAE has a deeper understanding of pixel-based text semantics and is more robust to different text forms, enabling it to classify all three cases correctly. In the fourth case, all three models perform the wrong classification. The words in the last image are minimal and blurry, and although humans can still distinguish some of the keywords, it is too difficult for the models.

From the above cases, we can find that OCR is not always so reliable and especially often fails for more complex document images. Our experimental results show that solving directly from pixels is a more direct and practical approach to understanding document content. Meanwhile, for more complex and fuzzy text, DocMAE still has room for improvement compared to human performance.

6 Conclusion

This paper investigated how to better understand the rich semantic content in document images. Given that the previous document image classification datasets mainly focused on document format while ignoring the document's text content, we presented a new dataset called DocCT. DocCT is the first dataset to concentrate on the topic classification of document images. The models must analyze fine-grained document content to classify each image under a correct topic. DocCT can facilitate the research related to document image understanding. Furthermore, we analyzed the shortcomings of previous document image classification models and presented a new self-supervised pretrained model called DocMAE. Compared to models that rely on OCR to obtain semantic text, DocMAE, as a purely pixel-based model, has better robustness, faster training and inference efficiency, and higher classification accuracy than previous methods on DocCT, proving it is possible to process document image semantics without OCR.

References

1. Antol, S., et al.: VQA: visual question answering. In: Proceedings of the IEEE International Conference on Computer Vision, pp. 2425–2433 (2015)
2. Appalaraju, S., Jasani, B., Kota, B.U., Xie, Y., Manmatha, R.: DocFormer: end-to-end transformer for document understanding (2021)
3. Bao, H., Dong, L., Wei, F.: BEiT: BERT pre-training of image transformers. arXiv preprint: arXiv:2106.08254 (2021)
4. Chen, N., Blostein, D.: A survey of document image classification: problem statement, classifier architecture and performance evaluation. IJDAR **10**(1), 1–16 (2007)
5. Chen, X., et al.: Context autoencoder for self-supervised representation learning. arXiv preprint: arXiv:2202.03026 (2022)
6. Deng, J., Dong, W., Socher, R., Li, L.J., Li, K., Fei-Fei, L.: ImageNet: a large-scale hierarchical image database. In: 2009 IEEE Conference on Computer Vision and Pattern Recognition, pp. 248–255. IEEE (2009)
7. Devlin, J., Chang, M.W., Lee, K., Toutanova, K.: BERT: pre-training of deep bidirectional transformers for language understanding. arXiv preprint: arXiv:1810.04805 (2018)
8. Dosovitskiy, A., Beyer, L., Kolesnikov, A., Weissenborn, D., Houlsby, N.: An image is worth 16x16 words: transformers for image recognition at scale (2020)
9. Harley, A.W., Ufkes, A., Derpanis, K.G.: Evaluation of deep convolutional nets for document image classification and retrieval. In: 2015 13th International Conference on Document Analysis and Recognition (ICDAR), pp. 991–995. IEEE (2015)
10. He, K., Chen, X., Xie, S., Li, Y., Dollár, P., Girshick, R.: Masked autoencoders are scalable vision learners. In: Proceedings of the IEEE/CVF Conference on Computer Vision and Pattern Recognition, pp. 16000–16009 (2022)
11. Huang, Y., Lv, T., Cui, L., Lu, Y., Wei, F.: LayoutlMv3: pre-training for document AI with unified text and image masking (2022)
12. Kim, G., et al.: OCR-free document understanding transformer. In: Avidan, S., Brostow, G., Cisse, M., Farinella, G.M., Hassner, T. (eds.) Computer Vision - ECCV 2022. LNCS, vol. 13688, pp. 498–517. Springer, Cham (2022)
13. Kingma, D.P., Ba, J.: Adam: a method for stochastic optimization. arXiv preprint: arXiv:1412.6980 (2014)
14. Lewis, D.D.: Building a test collection for complex document information processing. In: International ACM SIGIR Conference on Research and Development in Information Retrieval (2006)
15. Li, J., Xu, Y., Lv, T., Cui, L., Zhang, C., Wei, F.: DiT: self-supervised pre-training for document image transformer. arXiv preprint: arXiv:2203.02378 (2022)
16. Mathew, M., Karatzas, D., Jawahar, C.: DocVQA: a dataset for VQA on document images. In: Proceedings of the IEEE/CVF Winter Conference on Applications of Computer Vision, pp. 2200–2209 (2021)
17. Rolfe, J.T.: Discrete variational autoencoders. arXiv preprint: arXiv:1609.02200 (2016)
18. Vaswani, A., et al.: Attention is all you need. In: Advances in Neural Information Processing Systems, vol. 30 (2017)
19. Wang, J., Jin, L., Ding, K.: LiLT: a simple yet effective language-independent layout transformer for structured document understanding (2022)
20. Wang, Z., Shan, X., Zhang, X., Yang, J.: N24News: a new dataset for multimodal news classification. In: Proceedings of the Language Resources and Evaluation Conference, pp. 6768–6775. European Language Resources Association, Marseille, France (2022). https://aclanthology.org/2022.lrec-1.729

The Rach3 Dataset: Towards Data-Driven Analysis of Piano Performance Rehearsal

Carlos Eduardo Cancino-Chacón[✉][iD] and Ivan Pilkov[iD]

Institute of Computational Perception, Johannes Kepler University Linz,
Linz, Austria
carlos_eduardo.cancino_chacon@jku.at
http://www.carloscancinochacon.com

Abstract. Musicians spend more time practicing than performing live, but the process of rehearsal has been understudied. This paper introduces a dataset for using AI and machine learning to address this gap. The project observes the progression of pianists learning new repertoire over long periods of time by recording their rehearsals, generating a comprehensive multimodal dataset, the *Rach3 dataset*, with video, audio, and MIDI for computational analysis. This dataset will help investigating the way in which advanced students and professional classical musicians, particularly pianists, learn new music and develop their own expressive interpretations of a piece.

Keywords: Music Performance · Music Rehearsal · Data-driven Analysis · Multimodal dataset

1 Introduction

Musicians invest more time in practice than live performances or recording, yet the aspect of rehearsal remains largely understudied. This paper introduces the *Rach3* dataset, a multimodal dataset that enables a computational, data-driven approach to studying music rehearsal over long periods of time, leveraging current advancements in artificial intelligence and machine learning. This work focuses on piano music, as the piano is one of the most popular instruments and among the most extensively studied in music research [3,18,22].

The central questions of this large scale project concern the strategies musicians employ during rehearsal and the evaluation of performance quality in relation to the final product throughout the rehearsal phase. Addressing these questions is of practical and theoretical interest. Typically, music teachers assist students in evaluating their progress and identifying areas for improvement. Startups like Yousician[1] and Simply Piano[2] are increasingly interested in creating

---

[1] https://yousician.com.
[2] https://www.hellosimply.com/simply-piano.

This work is supported by the European Research Council (ERC) under the EU's Horizon 2020 research and innovation programme, grant agreement No. 101019375 (*Whither Music?*).

S. Rudinac et al. (Eds.): MMM 2024, LNCS 14565, pp. 28–41, 2024.
https://doi.org/10.1007/978-3-031-56435-2_3

automated music education tools. Yet, existing solutions are mostly limited to relatively simple aspects of music performance, like hitting the correct notes at approximately the right time. As of now, these tools are unable to help students develop efficient practice strategies or foster creative expressive interpretations.

On the theoretical side, studying the process of rehearsal can help us understand the way in which performers construct an expressive interpretation of the music. Musicologist Nicholas Cook very aptly comments that the role of performance within traditional musicology has been until recently downplayed, locating the *"aesthetic centre of the music in the written text"* [7]. Nevertheless, performance is a fundamental part of the musical experience. Through the study of the rehearsal process, we could investigate the way in which performers develop an understanding of the musical structure of the piece (e.g., syntactically and semantically meaningful segmentation and identifying which are the most salient elements at any given time), and shed some light into the way in which the procedural and implicit understanding of the musical structure developed by performers can be related to structural aspects in the music score.

Music rehearsal is an extended process, taking place over months or years, and involves skill development, evolving interpretations, and occasional periods of decline followed by re-learning [4,24]. To understand the rehearsal process, a comprehensive zoomed-out perspective is required, as well as the capability to zoom-in to examine individual rehearsal sessions [5,24]. But rehearsal as a longitudinal (and multimodal) process is difficult to capture: getting people to come and practice for 15 min in the lab is not sufficient, and just capturing audio or video is also insufficient.

This paper presents first results of the Rach3 dataset, a large-scale multimodal dataset containing audio, video and MIDI data[3] enabling cross-modal computational analysis of piano rehearsal. This dataset allows for a comprehensive and ecologically valid analysis of the rehearsal process over an extended period, which has been limited in previous research due to technical constraints and data availability (cf., the scale and scope of the studies by Chaffin and colleagues [4,5]). The central part of this dataset focuses on the first author's[4] learning of Rachmaninoff's challenging Piano Concerto No. 3 Op. 30 (commonly referred to as the Rach3 and hence the name of the dataset). A video playlist demonstrating potential analysis of this dataset using machine learning methods can be found at the following link.[5] Data collection started in Fall 2020 during a period of COVID lockdowns and travel restrictions, and there are currently around 350 h of recordings, which amounts to ca. 11 million performed notes (ca. 6TB). Due to the sheer size and complexity of the data, we have decided not to release the entire dataset all at once. Instead, we will make it accessible to the public in periodic subsections to ensure manageable access. The dataset will

---

[3] The Musical Instrument Digital Interface (MIDI) is a standardized protocol that enables digital musical instruments to communicate and record music.

[4] The first author is a professionally trained musician with a degree in Piano Performance and 14 years of formal music education.

[5] https://youtube.com/playlist?list=PLPUWmNCGflVOcjb5p4-ae3zFm0Z5l15RH.

be found, with detailed descriptions and regular updates, online at the following link.[6]

The rest of this paper is structured as follows: Section 2 discusses related work on music rehearsal research. Section 3 introduces the Rach3 dataset and its contents. Section 4 presents preliminary results of the dataset. Finally, Sect. 5 concludes the paper.

2   Related Work

Since most of the literature on music rehearsal comes from music psychology, in this section we aim to provide some context on the limitations of the scope in these approaches, motivating the need for the dataset presented in this work.

To properly explore research on music rehearsal, it is important to contextualize it within the broader field of performance research. The study of music performance spans various disciplines, including music psychology, musicology, and music information retrieval (MIR), each with its own focus. Music psychology investigates technical aspects like movement [16,26], as well as higher-level concepts such as emotion [17] and mental imagery [2]. Musicology has delved into historical performance practices [28] as well as examining the role of performance in understanding musical structure [6,7]. MIR and computational approaches have focused on analyzing the relationship between structural elements of the score and performance, both for analysis and generation purposes [3,18].

The study of music rehearsal gained momentum in the last few decades [21, 24]. A substantial number of studies focus on what musicians do when they practice, in particular on the decisions and goals that musicians take during the rehearsal process (e.g., deliberate practice, practice strategies and mental practice) [15,21].

One of the most influential works is Ericsson et al.'s research on the role of deliberate practice in order to attain expertise [10]. Work by Hallam et al. [13] and Chaffin et al. [4,5] examined the rehearsal process by observing music practice behaviors and exploring how approaches to rehearsing change over time via retrospective accounts.

However, music rehearsal remains understudied in a large data-driven way, and a lot of the research are case studies. In Miksza's review paper [21], none of the reviewed studies involved more than forty hours of recordings of rehearsals (see Table 1 in [21]), and the scope of their data and analysis was much smaller than the one proposed in this project. This is partly due to technical and logistical limitations in capturing long-term performance data, and the lack of efficient algorithms to extract relevant information and patterns from such a large source of data. Unlike common datasets used in computational approaches to studying expressive performance (e.g., ASAP [23] dataset), rehearsal data is fragmented, unstructured, and contains idiosyncratic variability and errors. This poses a challenge for computational approaches that assume performances are continuous or well-segmented.

---

[6] https://dataset.rach3project.com/.

Research on music psychology has emphasized the role of human movement in music performance through concepts like embodiment, which highlights the importance of the body in action, perception, and the mind-environment interaction [16]. Several studies have examined the roles of movements during music performance, investigated movement strategies of experts, and explored efforts to minimize effort and fatigue [8,11]. Additionally, research has focused on the differences in movement quality between experts and novices [8,29].

This type of research has been possible thanks to advancements in technologies such as optical motion capture. These technologies are however expensive and hard to bring into ecological settings, thus limiting this research to focused studies in lab settings. Automatic pose estimation [20], also known as markerless motion capture, is a machine learning-powered alternative for optical motion capture that estimates position of the body directly from video information. It has the advantage that it does not require physical markers, thus allowing for more naturalistic data capture. However, current methods have not been explicitly designed to deal with the intricacies of music playing.

3   The Rach3 Dataset

3.1   Desiderata for Datasets for Music Rehearsal Research

To ensure suitability for long-term, ecological studies of music rehearsals, a dataset must fulfill several criteria. Firstly, it should be *representative* of typical rehearsal processes, capturing pianists' behavior that closely resembles regular rehearsals. Secondly, it should aim for *ecological validity* by capturing the conditions of normal rehearsals as closely as possible, including location and rehearsal structure. Lastly, it should be *comprehensive in scope*, incorporating diverse data sources for both quantitative and qualitative analysis. The Rach3 dataset has been developed to meet these requirements, implementing five specific points. The collected data already possesses these characteristics, and ongoing efforts will maintain these standards throughout the data collection process.

1. **Multimodality.** A comprehensive study of music performance requires both visual and auditory components [8,27]. The process of learning a piece involves both the musical and motor aspects from an embodied music perspective. Therefore, a rehearsal dataset should include video and audio recordings. For piano performance, precise measurements of when and how each note is played are necessary to study patterns of variability in expressive parameters like timing and articulation [12]. Extracting played notes from audio alone is a complex task and a focus of research in MIR [1]. To address this issue, recordings in the Rach3 dataset are made on acoustic pianos equipped with sensors that detect key movement and convert it to MIDI signals, such as Yamaha's Silent Pianos[7] or Disklaviers.[8]

---

[7] https://usa.yamaha.com/products/musical_instruments/pianos/silent_piano/index.html.

[8] https://usa.yamaha.com/products/musical_instruments/pianos/disklavier/.

2. **Multiple time scales.** The dataset should cover extended time periods, with a minimum of 12 months and over 200 h of recordings for each pianist. It should include regular rehearsal sessions without significant gaps between recordings. This comprehensive coverage enables the study of a piece's evolution within and across rehearsal sessions, as well as the detailed monitoring of development of its expressive interpretation. At the time of this writing, there have not been gaps between recordings longer than 4 weeks (these gaps have been due to traveling overseas for holidays or conferences).

3. **Repertoire.** The selection of repertoire should have two main focuses: the process of learning new pieces from scratch (starting from sight reading and progressing to performance readiness), and the regular maintenance of previously learned repertoire. This will enable comparisons between rehearsal strategies used when learning a piece for the first time, techniques for keeping a piece in good shape, and the process of relearning music that has been previously studied. A central component of the proposed dataset focuses on the first author's rehearsals for both practical reasons such as the need to pilot test the recording setup and data processing methods; as well as the opportunity for introspection, allowing for hypothesis development and understanding of the rehearsal process. This introspection is crucial in the exploratory stage, given the limited existing research (note also that the Chaffin et al. studies are partly introspective, since the musicians are co-authors of the studies [5,9] and see point 5 below). Rachmaninoff's Third Piano Concerto was chosen as the focus of rehearsal due to its technical challenges and reputation within classical piano repertoire. Selecting this piece minimizes self-study bias, since it is challenging enough that it leaves little room for deviating from normal rehearsal strategies. The piece showcases various technical challenges, such as intricate fast passages with complex rhythms, demanding hand extensions for large chords, and alternating between rapid and cantabile sections. The rehearsals in the dataset include not only Rachmaninoff's concerto, but also learning other pieces from scratch as well as practicing/re-learning pieces played before.

4. **Unobtrusive data capturing.** The dataset should strive to capture the authenticity of real-life rehearsal situations, considering disruptions that may occur such as relocation, different pianos, and stress, as well as performance distractions like interruptions from pets or external noises. Additionally, the recordings should be as unobtrusive as possible, minimizing any impact on the pianist's movements or potential distractions. To achieve this, the setup for recording should be quick and efficient, taking only a few minutes, and avoiding the use of sensors that could interfere with the pianist's performance.

5. **Self-reflection and documentation.** Each rehearsal session should be accompanied by a documented reflection from the pianists, in which they discuss the specific focus and goals of the different parts of the rehearsal.

3.2   Dataset Contents

The Rach3 dataset consists of the following components:

1. **Multimodal rehearsal recordings.** All rehearsal sessions are captured with synchronized video, audio and MIDI, allowing for high-precision piano note-level recording. Rehearsals have mostly taken place on three pianos: a Yamaha GB1k Silent, an Essex EUP-116E with a built-in silent system, and a Yamaha C1 Disklavier Enspire ST.[9] To record video, a video camera (GoPro) is placed above the keyboard, capturing the position/movement of the hands (including fingers), arms, head and torso (see Fig. 5). Video is recorded at 1080p at 60 frames per second (fps).[10] Audio is recorded at 44.1 kHz using condenser microphones (2 AKG P120, AKG P170 or AKG P240, depending on the location) connected to a USB audio interface (Focusrite Scarlett 2i2).

2. **Rehearsal documentation and subjective experience data.** A rehearsal log chronicles the focus of each rehearsal session, and the Multidimensional Mood State Questionnaire (MDBF) [25] is employed to assess the pianists' mood after each session. We will utilize the MDBF questionnaire and the rehearsal log to contextualize day-to-day variations in rehearsal strategies. The rehearsal log serves a dual purpose: it documents played pieces or sections for labeling recordings and tracks rehearsal focus, such as strategies or goals, for qualitative analysis, in a similar way to the approaches used by Roger Chaffin's musician colleagues Gabriela Imreh and Tânia Lisboa [4,5].

3. **Machine-readable music scores.** The dataset will also incorporate machine-readable music scores (in formats like MusicXML or MEI) for all pieces. For practical reasons, we will ask all contributing pianists to concentrate on music from the Common Practice Period,[11] as these pieces are in the public domain and machine-readable scores may already be available online in repositories such as IMSLP[12] or MuseScore.[13] If a piece's score is unavailable, we will semi-automatically create it using optical music recognition and manual correction.

3.3 Synchronization of Multimodal Data

Audio and MIDI signals are captured using a digital audio workstation (DAW) to ensure synchronization. The main challenges in aligning video and audio/MIDI are differences in their start and end times and potential drift, which might arise from minor recording delays or desynchronization between devices' internal clocks. Instead of altering the speed or duration of recordings, we synchronize video and audio by aligning the audio spectrogram of the audio track of the video

---

[9] When recording rehearsals on silent pianos, it is possible to just get the output MIDI signal from the optical sensors built into the piano, without mechanically silencing the piano, which enables capturing MIDI and the real sound of the piano.

[10] Early in the dataset, recordings were made at 30 fps, and a few at 120 fps. Higher frame rates caused the GoPro to shut down due to overheating, so 60 fps was chosen as a compromise between higher resolution and long recording times.

[11] This period corresponds roughly to the Baroque, Classical, Romantic, and early 20th Century periods of Western Classical music.

[12] https://imslp.org/wiki/Main_Page.

[13] https://musescore.com.

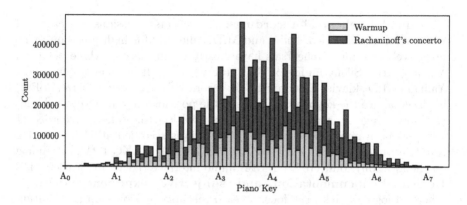

Fig. 1. Histogram of notes performed in the Rach3 dataset

and the recorded audio track. This involves calculating magnitude spectrograms for both the video's audio track and the recorded audio. Given our large dataset, full alignments using dynamic time warping are impractical. Instead, we use the MIDI file to identify the onset time of the first note, and compare the audio signals' spectrograms around this time using cosine distance. To counter drift, recordings are segmented based on pauses in the MIDI file (up to a pre-defined maximum duration), and each segment undergoes the same spectrogram synchronization.

4   Preliminary Results

4.1   Dataset Statistics

Data collection started in mid August 2020, and at the time of this writing, there are nearly 400 recorded rehearsal sessions, each with an average duration of 53 min. 56% of rehearsals have been recorded on the Yamaha GB1k, 39% on the Essex EUP-116E and 5% on the Yamaha Disklavier. A typical rehearsal consists of a warmup consisting of technical exercises (e.g., Hanon exercises, scales) and other pieces. The warmup is then followed by practicing Rachmaninoff's concerto, and nearly two thirds of the time has been dedicated to the concerto. Figure 1 shows the distribution of played notes in the dataset. In this figure, we can see that the most frequent notes are D_4, D_5, A_4, and F_4, which belong to the key of D minor, which is the key of Rachmaninoff's piano concerto.

The Rach3 dataset currently boasts 350 h of recordings, making it the most extensive collection of synchronized piano recordings and MIDI to date. It surpasses the MAESTRO dataset [14], previously the largest of its kind, by almost 75% (see Table 1). Furthermore, it's substantially more extensive than both the Magaloff and Batik datasets, designed for the computational modeling of expressive performance [3].

Table 1. Size comparison of datasets for music performance research.

	Rach3			MAESTRO				Magaloff	Batik
	Warmup	Concerto	Total	Train	Val	Test	Total		
Notes (millions)	3.46	7.17	**10.62**	5.66	0.64	0.74	**7.04**	**0.30**	**0.10**
Duration (hours)	123.4	226.0	**349.4**	159.2	19.4	20.0	**198.7**	**10.0**	**3.0**

Fig. 2. Word cloud of terms in the rehearsal log.

By the project's conclusion, our goal is to gather over 1000 h of practice sessions from at least four different pianists (see Sect. 5). It's worth noting that as of now, we have already collected 60+ hours from a second professionally trained pianist, with a performance degree from the Royal Conservatory of Music in Toronto, Canada, and 16 years of formal musical education. However, as these recordings are unprocessed, they're not factored into the statistics presented in this section.

The rehearsal log contains around 46.6k words, of which around 11% are unique. A word cloud of the rehearsal log is shown in Fig. 2. In this figure we can see that many of the most prominent terms correspond to description of parts of the piece (e.g., *"first movement"*, *"theme B"*, *"bar"*, *"section"*), with other terms focusing on how did the performance go (e.g., *"cleaning"*, *"dirty"*, *"hiccup"*, *"good"*).

We utilize a pretrained roBERTa model from TimeLMs [19] to determine if the comments in the log for each rehearsal session convey a positive or negative sentiment. The confidence score from the model, which indicates the certainty of a sentiment being positive/negative/neutral, serves as our sentiment score. A positive sentiment multiplies this score by 1, whereas a negative sentiment multiplies it by −1. If the sentiment is neutral, we apply a second model, a pretrained DistilBert fine-tuned on the SST-2 dataset,[14] to ascertain whether the comment is positive or negative. We then use the original roBERTa score to

---

[14] https://huggingface.co/distilbert-base-uncased-finetuned-sst-2-english.

reflect sentiment intensity. Figure 3 presents the sentiment progression throughout the rehearsals. The graph suggests an increasing trend of positive comments
over time. This shift could be influenced by numerous factors, such as environmental elements like the COVID pandemic and the 2020-2021 lockdowns, or due
to enhanced proficiency in the rehearsed pieces. Figure 4 shows an example of
positive and negative comments, and their respective position in the musical
score.

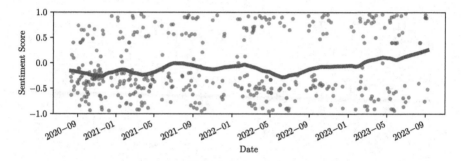

Fig. 3. Sentiment score for the comments of the rehearsal log across the rehearsal
period. The dots represent the predictions of the model and the red line corresponds
to the trend. (Color figure online)

"Finally solved the broken-sixteenths passage in bars
140-143!"

"Focused on cleaning Section 1, since it has some issues. It
feels that every time that something is getting better, then
the rest starts decaying and rotting. I had to spend a good
chunk of time cleaning the sixteenth runs (unison octaves
in bars 54-57 and 59-61), which have been quite dirty for
the last few days. This is not the most fun part of practicing
the piece. I would have like to play more, but since I had so
many things to do, I had to move to the third movement.
Section 11 is now more fluent, but is still very slow. I feel
that my playing is a bit heavy."

Positive Comment   **Negative Comment**

Fig. 4. Example of positive and negative comments

Table 2. Estimated quantities of motion for three sample performances. The distance traveled and the average velocity are measured in the image plane (approx. parallel to the musical keyboard).

Piece	Duration(s)	Keypoint	Distance(m)	Avg.Velocity(m/s)
Hanon's Piano Exercise 1	37.45	Right wrist	2.04	0.05
		Left wrist	1.94	0.05
Mozart's Variations K 265	97.05	Right wrist	8.52	0.09
		Left wrist	7.58	0.08
Rachmaninoff's Opus 30	353.40	Right wrist	50.84	0.14
		Left wrist	71.57	0.20

4.2 Motion Analysis

An important application of the dataset is the analysis of human motion during a musical performance. We conducted some preliminary experiments studying the feasability of such an analysis using Google's pretrained automatic hand pose estimation model from MediaPipe [30] in order to obtain the relative 3D spatial coordinates of the hands and fingers of the performer from the video recordings. This two-stage model uses a palm detector followed by a hand landmark model to obtain 21 landmarks with 3 coordinates each per single hand for each individual video frame (shown in Fig. 5). Therefore we can obtain estimated movement quantities for a performance such as total traveled distance and average velocity, as shown in Table 2 for a few selected performances and hand keypoints. We consider the wrist position as an indication of the movement of the hand across the piano keyboard. From the results in Table 2, we can already observe differences in this quantity between both the right and the left hand (resulting from bigger musical intervals and arpeggios in the rhythmic part versus smaller intervals in the melodic part) and between different performances as well. The latter correlates with and could be used as an indication of the complexity of a piece (as measured by the note density, the inter-onset intervals, the tempo, and the general spread of notes across the piano keyboard), but also depends on the expressive characteristics of each individual performance as well as the acquired piece memorization and efficiency of hand movements.

Further analysis could be done by creating motion videos from subsequent frame differences and obtaining different representations as for example motiongrams [16], which would allow a visual study of efficiency of movement during a performance as well as a comparative study of different performances of the same piece, while also serving as an entry into more comprehensive musical gestures research. A small example is given in Fig. 6 where such a motiongram of a short excerpt of a performance of Mozart's Variations on "Ah vous dirai-je, Maman" K 265 is shown next to a plot of the corresponding piano roll, which

Fig. 5. Automatic hand pose estimation results with a missing detection of the right hand in the middle frame.

showcases the note distribution over the keyboard against the spatial position of both hands' movement patterns over time.

However, before more conclusive analysis in these and similar regards could be conducted, a number of issues would need to be resolved. Currently the hand pose estimation model used is not yet robust enough to deal with fast passages and sudden movements of the hands and it often has problems with the initial hand detection in changing lighting conditions (Fig. 5). An additional hand tracking model to predict the next frame position of each hand could bypass the need to utilize the computationally costly first stage palm detector for each frame of a video, while an improvement of the palm detector by utilizing a domain adapta-

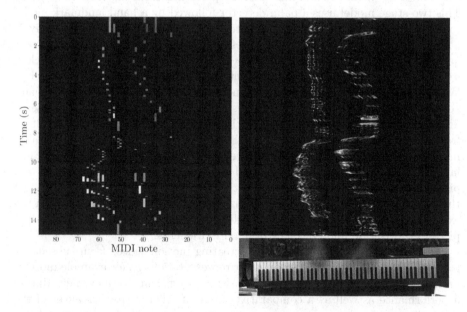

Fig. 6. Example of motion analysis on a 15 s excerpt of a recording of Mozart's Variations on "Ah vous dirai-je, Maman" K 265. Left is a piano roll of the played notes, right is a vertical motiongram with the average image of the sequence below.

tion for our application would result in higher initial detection percentage. Similarly, using smoothing filters could potentially minimize landmarks' positional fluctuation, which currently makes distance estimation especially unreliable for individual finger keypoints. To validate (and improve) the automatic pose estimation models, we are gathering a smaller piano performance dataset with the Rach3's recording setup plus optical motion capture to serve as ground truth for the motion analysis. Finally, to deal with the inherently costly video processing needed for motion analysis, a number of additional considerations are: using estimates for movement quantities by averaging, sacrificing spatial and/or temporal precision by compressing videos, running model inference on non-consecutive frames and interpolating, and so on.

5 Conclusions and Future Work

This paper introduced and presented first results of the Rach3 Dataset, part of an ongoing project to study performance rehearsal. Future work will focus on the following points:

1. **Data Cleaning and Publishing the Dataset**. Besides the recordings themselves, a semi-automatic data cleaning and annotation phase including synchronization of video with audio and MIDI, identifying which pieces/sections are being played in each recording, as well as mapping performance strategies (from the rehearsal log) to time points in the recording (e.g., practicing a passage staccato/legato). The data cleaning will use a mix of MIR and machine learning methods as well as manual annotations. Besides the cleaning, there are a number of practical issues concerning the immediate publication of the Rach3 Dataset, most important of which is the ongoing data collection process. Since this is expected to continue for at least the next few years, we plan to publish smaller subsets of the collected data in regular intervals. This would allow access to researchers for early experiments, while continuing the long-term rehearsal research going uninterrupted.

2. **Rehearsals by Other Pianists**. In order to diversify the dataset we plan to include recordings by different pianists from different levels of expertise. Besides the current recordings of the first author's performances, we aim to include additional recordings by two university-enrolled piano students (a first year Bachelor's student and an advanced Bachelor's/Master student), as well as recordings by another classically trained pianist with a degree in piano performance, estimated at 200+ hours per pianist. The recordings by this last pianist started in Fall 2022 and there already are more than 60 recorded rehearsal sessions. All additional recordings will use the same recording setup to that of Rachmaninoff's piano concerto rehearsals. None of the other pianists will receive any input on how they should organize their practice or select pieces and each pianist will fill their own rehearsal logs and MDBF questionnaires at the end of each of their respective sessions.

References

1. Benetos, E., Dixon, S., Duan, Z., Ewert, S.: Automatic music transcription: an overview. IEEE Signal Process. Mag. **36**(1), 20–30 (2019). https://doi.org/10. 1109/MSP.2018.2869928
2. Bishop, L., Bailes, F., Dean, R.T.: Performing musical dynamics. Music. Percept. **32**(1), 51–66 (2014). https://doi.org/10.1525/mp.2014.32.1.51
3. Cancino-Chacón, C.E., Grachten, M., Goebl, W., Widmer, G.: Computational models of expressive music performance: a comprehensive and critical review. Front. Digital Human. **5**, 25 (2018). https://doi.org/10.3389/fdigh.2018.00025
4. Chaffin, R., Imreh, G.: Practicing perfection: piano performance as expert memory. Psychol. Sci. **13**(4), 342–349 (2005). https://doi.org/10.4324/9781410612373
5. Chaffin, R., Lisboa, T., Logan, T., Begosh, K.T.: Preparing for memorized cello performance: the role of performance cues. Psychol. Music **38**(1), 3–30 (2010). https://doi.org/10.1177/0305735608100377
6. Cook, N.: Analysing performance and performing analysis. In: Cook, N., Everist, M. (eds.) Rethinking Music, pp. 239–261. Oxford University Press, Oxford (1999)
7. Cook, N., Johnson, P., Zender, H.: Theory into Practice: Composition, Performance And The Listening Experience. Leuven University Press (2021). https://doi.org/ 10.2307/j.ctv1rh36q7
8. Dahl, et al.: Gestures in Performance. In: Musical Gestures: Sound, Movement, and Meaning, pp. 36–68. Routledge (2010)
9. Demos, A.P., Lisboa, T., Chaffin, R.: Flexibility of expressive timing in repeated musical performances. Front. Psychol. **7**, 1490 (2016). https://doi.org/10.3389/ fpsyg.2016.01490
10. Ericsson, K.A., Krampe, R.T., Tesch-Romer, C.: The role of deliberate practice in the acquisition of expert performance. Psychol. Rev. **100**(3), 364–403 (1993)
11. Goebl, W.: Movement and touch in piano performance. In: Müller, B., Wolf, S.I., Brueggemann, G.P., Deng, Z., McIntosh, A., Miller, F., Selbie, W.S. (eds.) Handbook of Human Motion. Springer International Publishing, Cham (2017). https:// doi.org/10.1007/978-3-319-30808-1
12. Goebl, W., Dixon, S., Poli, G.D., Friberg, A., Widmer, G.: Sense in expressive music performance: data acquisition, computational studies, and models. In: Polotti, P., Rocchesso, D. (eds.) Sound to Sense - Sense to Sound: A State of the Art in Sound and Music Computing, pp. 195–242. Logos, Berlin (2008)
13. Hallam, S., Papageorgi, I., Varvarigou, M., Creech, A.: Relationships between practice, motivation, and examination outcomes. Psychol. Music **49**(1), 3–20 (2021). https://doi.org/10.1177/0305735618816168
14. Hawthorne, C., et al.: Enabling Factorized Piano Music Modeling and Generation with the MAESTRO Dataset. In: Proceedings of the International Conference on Learning Representation, New Orleans, USA (2019)
15. How, E.R., Tan, L., Miksza, P.: A PRISMA review of research on music practice. Musicae Scientiae **26**(3), 455–697 (2022). https://doi-org.ezproxy.uio.no/10.1177/ 10298649211005531
16. Jensenius, A.R., Wanderley, M.M., Godøy, R.I., Leman, M.: Musical gestures concepts and methods in research. In: Godøy, R.I., Leman, M. (eds.) Musical Gestures: Sound, Movement, and Meaning, pp. 12–35. Routledge (2010)
17. Juslin, P.N., Laukka, P.: Communication of emotions in vocal expression and music performance: different channels, same code? Psychol. Bull. **129**(5), 770–814 (2003). https://doi.org/10.1037/0033-2909.129.5.770

18. Lerch, A., Arthur, C., Pati, A., Gururani, S.: An interdisciplinary review of music performance analysis. Trans. Int. Soc. Music Inf. Retr. **3**(1), 221–245 (2020). https://doi.org/10.5334/tismir.53

19. Loureiro, D., Barbieri, F., Neves, L., Espinosa Anke, L., Camacho-Collados, J.: TimeLMs: diachronic language models from Twitter. In: Proceedings of the 60th Annual Meeting of the Association for Computational Linguistics: System Demonstrations, pp. 251–260. Dublin, Ireland (2022). https://doi.org/10.18653/v1/2022.acl-demo.25

20. Mathis, A., Schneider, S., Lauer, J., Mathis, M.W.: A primer on motion capture with deep learning: principles, pitfalls, and perspectives. Neuron **108**(1), 44–65 (2020). https://doi.org/10.1016/j.neuron.2020.09.017

21. Miksza, P.: A review of research on practicing: summary and synthesis of the extant research with implications for a new theoretical orientation. Bull. Counc. Res. Music. Educ. **190**, 51–92 (2011). https://doi.org/10.5406/bulcouresmusedu.190.0051

22. Palmer, C.: Music performance. Annu. Rev. Psychol. **48**, 115–138 (1997)

23. Peter, S.D., et al.: Automatic note-level score-to-performance alignments in the ASAP dataset. Trans. Int. Soc. Music Inf. Retr. **6**(1), 27–42 (2023). https://doi.org/10.5334/tismir.149

24. Reid, S.: Preparing for performance. In: Rink, J. (ed.) Musical Performance, pp. 102–112. Cambridge University Press (2002). https://doi.org/10.1017/CBO9780511811739.008

25. Steyer, R., Schwenkmezger, P., Notz, P., Eid, M.: Development of the Multidimensional Mood State Questionnaire (MDBF). Primary data. (Version 1.0.0) [Data and Documentation]. Tech. rep., Trier: Center for Research Data in Psychology: PsychData of the Leibniz Institute for Psychology ZPID (2004)

26. Verdugo, F., Pelletier, J., Michaud, B., Traube, C., Begon, M.: Effects of trunk motion, touch, and articulation on upper-limb velocities and on joint contribution to endpoint velocities during the production of loud piano tones. Front. Psychol. **11**, 1159 (2020). https://doi.org/10.3389/fpsyg.2020.01159

27. Visi, F.G., Östersjö, S., Ek, R., Röijezon, U.: Method development for multimodal data corpus analysis of expressive instrumental music performance. Front. Psychol. **11**, 576751 (2020). https://doi.org/10.3389/fpsyg.2020.576751

28. Walls, P.: Historical performance and the modern performer. In: Rink, J. (ed.) Musical Performance, pp. 17–34. Cambridge University Press (2002). https://doi.org/10.1017/CBO9780511811739.003

29. Winges, S., Furuya, S.: Distinct digit kinematics by professional and amateur pianists. Neuroscience **284**, 643–652 (2015). https://doi.org/10.1016/j.neuroscience.2014.10.041

30. Zhang, F., et al.: MediaPipe hands: on-device real-time hand tracking. In: Proceedings of the CVPR Workshop on Computer Vision for Augmented and Virtual Reality, Seattle, WA, USA (2020)

WikiMuTe: A Web-Sourced Dataset of Semantic Descriptions for Music Audio

Benno Weck[1,2]($\boxtimes$) ⓘ, Holger Kirchhoff[1] ⓘ, Peter Grosche[1] ⓘ,
and Xavier Serra[2] ⓘ

[1] Huawei Technologies, Munich Research Center, Munich, Germany
{benno.weck,holger.kirchhoff,peter.grosche}@huawei.com
[2] Universitat Pompeu Fabra, Music Technology Group, Barcelona, Spain
benno.weck01@estudiant.upf.edu, xavier.serra@upf.edu

Abstract. Multi-modal deep learning techniques for matching free-form text with music have shown promising results in the field of Music Information Retrieval (MIR). Prior work is often based on large proprietary data while publicly available datasets are few and small in size. In this study, we present *WikiMuTe*, a new and open dataset containing rich semantic descriptions of music. The data is sourced from Wikipedia's rich catalogue of articles covering musical works. Using a dedicated text-mining pipeline, we extract both long and short-form descriptions covering a wide range of topics related to music content such as genre, style, mood, instrumentation, and tempo. To show the use of this data, we train a model that jointly learns text and audio representations. The model is evaluated on two tasks: tag-based music retrieval and music auto-tagging. The results show that while our approach has state-of-the-art performance on multiple tasks, we still observe a difference in performance depending on the data used for training.

Keywords: music information retrieval · cross-modal · text-mining

1 Introduction

Music is a complex and multi-faceted art form. Descriptions of music can therefore be very diverse, covering not only content-related aspects such as instrumentation, lyrics, mood, or specific music-theoretical attributes, but also contextual information related to artist or genre. With unprecedented access to vast amounts of music recordings across a multitude of genres and styles, the ability to search for music based on free-form descriptions has become increasingly important in recent years. Multi-modal deep learning techniques that match textual descriptions with music recordings have seen growing interest [8,9,12,13,17,20]. These methods aim to draw connections between semantic textual descriptions and the content of a music recording and have potential applications in text-to-music retrieval and text-based music generation. However, despite promising results, this area of research is still in its infancy, and the lack of suitable and open text-music datasets hinders progress in this field.

S. Rudinac et al. (Eds.): MMM 2024, LNCS 14565, pp. 42–56, 2024.
https://doi.org/10.1007/978-3-031-56435-2_4

Historically, text-music datasets were limited by the number of available text labels and researchers have turned to various data sources. For example, the use of short labels (tags) in automatic tagging systems has been studied extensively [3,21]. These labels are typically taken from online platforms, so-called *social-tags* [14]. However, due to data sparsity, researchers often only use a limited number of tags (e.g., top 50), which can cover only certain aspects of the music. This highlights the need for a more complex form of description that can capture the nuances of music and provide a richer set of labels for music retrieval and analysis.

To obtain free-form textual descriptions of music, researchers have sought other forms of supervision beyond social-tags. Some studies have used metadata from online music videos [12] or production music [17]. Since these datasets are not publicly available, different initiatives have proposed to create music description texts manually, either through crowd-sourcing [18,19] or through expert labelling [1]. While both methods can be slow and require significant resources, others have proposed to synthesise text descriptions, for example, by combining tags into longer text [9] or by employing large-language models (LLMs) for automatic generation [8,13,20]. While this is a cost-effective method to generate large text corpora, it brings the risk that the LLM produces inaccurate descriptions.

Research on multi-modal training in other areas such as computer vision or machine listening has benefited greatly from the availability of large-scale datasets. These datasets are often built by crawling the web for suitable data (e.g., images [23,28] or sound recordings [36]) paired with natural language texts and may contain millions of data points. Even though web mining to collect tags for music is common practice in MIR research [30], efforts for longer text-music descriptions are largely missing to date. Previous efforts to build multi-modal (music and text) datasets from online sources collected data at the artist or album level [22,26]. We are, however, interested in descriptions of music at a more granular level (i.e., song or segment level).

From these various forms of data sourcing, we identify that (i) there is a lack of openly accessible datasets (ii) manual labelling is laborious and cannot easily scale to large datasets (iii) synthesised texts might not reflect the resourcefulness of human descriptions.

To fill this gap, we present *WikiMuTe*, a new publicly available[1] web-sourced dataset of textual descriptions for music collected from encyclopedic articles in Wikipedia. Using a text-mining pipeline, we extract semantically and syntactically rich text snippets from the articles and associate those with corresponding audio samples. A cross-modal filtering step is subsequently employed to remove less relevant text-audio pairs. This approach allows us to collect a large amount of data describing music content (e.g., genre, style, mood, instrumentation, or tempo) which can serve as a suitable dataset for training deep learning models for text-to-music matching. The collected data was utilised in a series of experiments to demonstrate its practical value. These experiments involved text-to-music retrieval, music classification, and auto-tagging.

---

[1] https://doi.org/10.5281/zenodo.10223363.

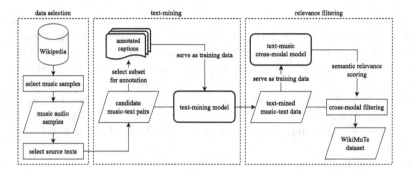

Fig. 1. Flowchart of our data mining pipeline

2 Dataset Compilation

We have chosen the online encyclopedia Wikipedia as our source of data since it contains a multitude of articles on music pieces and popular songs that are linked to corresponding audio samples of music recordings. Additionally, its content is freely accessible. To extract a high-quality dataset of music audio paired with textual descriptions, we propose a web mining pipeline consisting of three stages. First, we collect music samples and corresponding text from Wikipedia. Second, since the text covers a wide variety of information about the sample, we extract individual aspects and sentences that refer to descriptions of the music content thereby discarding non-content-related parts. This is achieved semi-automatically through a dedicated text-mining system. Third, we rate the semantic relevance of all music-text pairs and discard texts with low relevance. Figure 1 gives an overview of the steps involved. In the following sections, each of the three stages – data selection, text-mining and relevance filtering – is explained in more detail.

2.1 Source Data Selection

We select audio samples that are linked to the 'Music' category on Wikipedia and Wikimedia Commons, effectively excluding samples of speech, natural sounds, etc. The samples are associated with multiple potential text sources. We consider caption texts, file descriptions, and article texts. We only take article texts into account that are linked to Wikipedia's 'Songs' category. To avoid false matches, we exclude certain sections from the article using a manually defined list of headings, such as 'Music video', 'Chart performance', 'Covers' or 'Remixes'. Moreover, categorical information such as genre or instrument names is extracted from the metadata if available using simple heuristics. An example of an audio sample and caption is given in Fig. 2.

The collected texts are not readily usable in their raw form. They frequently contain information that is not pertinent to the audio content, thus introducing irrelevant data that can potentially skew the results of any analysis or application. Moreover, they present a challenge due to their excessive length.

"Take On Me"

A 23-second sample from "Take On Me", featuring Harket's high-pitched falsetto, with a backing track that mixes acoustic and electric guitars and electronic instrumentation

Fig. 2. Example of an audio sample with a caption on a Wikipedia page [33].

2.2 Text-Mining System

To automatically extract music descriptions from the larger bodies of text collected in the first stage, we construct a text-mining system. This system should be able to handle the complexity of music descriptions and accurately extract relevant information. Pretrained Transformer models are a state-of-the-art solution for this task and have shown promising results in text mining for other fields [11]. However, we need to fine-tune them to extract music descriptions.

Data Annotation. To obtain training data for our text-mining system, we manually labelled a subset of the previously collected text items. Specifically, we have selected the caption text of the associated audio sample as our annotation source. Caption texts are short and likely to contain content descriptions of the music recording.

In our annotations, we focus on descriptions of the musical content only. Context-dependent descriptions on the other hand, such as lyrics, forms of appraisal, and references to other artists or recordings are deliberately excluded as they cannot easily be matched with the content of the music sample. We annotated spans of the text in two steps: First, we annotated longer phrases such as full sentences and clauses, which we refer to as *sentences*. Second, we annotated short phrases and single words such as adjectives, nouns, and verbs, which we refer to as *aspects*. The former may be described as music captions or long-form descriptions, and the latter as tags or short-form. The two types of annotations may overlap, as aspects are often part of a longer phrase.

Text Mining and Data Collection. Using the annotated data, we train a system to detect sentences and aspects, respectively. Both systems are trained by fine-tuning a pretrained DistilBERT [7,25] model on a binary token classification task. This results in automatically extracted spans of text referring to aspects or sentences. We note that this extracted data is still quite noisy. It can contain descriptions that are not relevant to the music sample, for example, if the audio represents only a specific part of the song.

Table 1. Two examples of the collected data (RF = Relevance filtering)

Dave Brubeck Quartet - Take Five	
Text-mined aspects	'West Coast cool jazz', 'inventive', 'cool-jazz', 'quintuple', 'piano', 'saxophone', 'bass vamp', 'jazz', 'two-chord', 'roots reggae', 'jolting drum solo', 'scat'
Text-mined sentences	'is known for its distinctive two-chord piano/bass vamp; catchy, cool-jazz saxophone melodies; inventive, jolting drum solo; and unorthodox quintuple', 'the song has a moderately fast tempo of 176 beats per minute', 'is a jazz standard'
Removed by RF	'jolting drum solo', 'piano', 'roots reggae', 'the song has a moderately fast tempo of 176 beats per minute'
Pharrell Williams - Happy	
Text-mined aspects	'handclaps', 'faux-', 'falsetto voice', 'lead vocals', 'Soul', 'programmed drums', 'neo soul', 'sings', 'uptempo', '156 beats per minute', 'neosoul funk groove', 'Motown', 'falsetto', 'backing vocals', 'prightly', 'mid-tempo', 'bass', 'sparse', 'singing', 'vocal', 'soul', 'keyboard part'
Text-mined sentences	'The song is written in the key of F minor', 'and at a tempo of 156 beats per minute', 'is an uptempo soul and neo soul song on which', 'with a "sprightly neosoul funk groove"', 'falsetto voice', 'Williams sings the upper notes in falsetto', 'is "a mid-tempo ... song in a faux-Motown style, with an arrangement that is, by modern standards, very sparse: programmed drums, one bass and one keyboard part, and handclaps both programmed and played, all topped off by Williams's lead vocals and a whole posse of backing vocals"', 'is a mid-tempo soul and neo soul song'
Removed by RF	'sparse', 'faux-', 'uptempo', 'prightly'; 'and at a tempo of 156 beats per minute'

2.3 Semantic Relevance Filtering

To ensure that audio and text are matching, we experiment with cross-modal filtering. Prior work has applied this form of filtering using multi-modal deep learning models as a means to select suitable training data [13,23]. By employing models that are trained to give a high score to semantically related data points and a low score to non-matching data points, irrelevant text data can be filtered. We adopt this idea and employ a model for text-to-audio alignment as a scoring function of how well a description matches a music piece.

To test if the text-mined data are a fitting description for their respective audio sample, we pair each text item with its respective audio sample and compute their score. For each audio sample that exceeds the input length expected by the model, we compute multiple scores by segmenting the sample into non-overlapping blocks. The final relevance score is computed by taking the average. In our implementation, the score is provided as the cosine similarity between the vector representations (embeddings) of audio and text. Finally, we remove all text items with a negative cosine similarity value.

Table 2. Statistics of text-music datasets

Dataset	Tracks	Item duration	Text source	Vocab. size	Public text	Public audio
WikiMuTe	9 k	snippets & full-length tracks	text-mining	14 392	✓	✓
MusicCaps [1]	5.5 k	10 s	hand-labelled	7578	✓	✗
LP-MusicCaps [8]	500 k	30 s	LLM generated	39 756	✓	✗
eCALS [9]	500 k	30 s	social tags	1054	✓	✗
MusCall [17]	250 k	full-length tracks	metadata	-	✗	✗
MULAN [12]	40 M	10 s	metadata	-	✗	✗

2.4 The WikiMuTe Dataset

The final dataset contains free-form text descriptions for 9000 audio recordings. On average the tracks have a duration of 81 s (Median: 29 s). The descriptions cover a wide range of topics related to music content such as genre, style, mood, instrumentation, and tempo. Also, they can include information on the era, production style, and even individual sounds. Table 1 gives two examples of the extracted texts. Due to its richness, we posit that the data can be used as a general-purpose music-text dataset and is applicable to several use cases such as cross-modal retrieval and music generation.

Table 2 compares WikiMuTe with other datasets used in contrastive learning studies for text-to-music matching. Out of these datasets, the one that stands out as most relevant for comparison is *MusicCaps* [1] due to its size and content, which are comparable to our dataset. MusicCaps is derived from the AudioSet dataset [10] and contains 5500 ten-second audio clips labelled with multiple descriptions in the form of a *caption* and an *aspect* list. It was constructed for music generation from text prompts. We list the top ten most common aspects in Table 3. On average, WikiMuTe contains 7.6 (Median: 5) aspects per recording and MusicCaps 10.7 (Median: 9); respectively there are 2.5 (Median: 1) and 3.9 (Median: 4) sentences.

Despite aiming for the inclusion of all kinds of music descriptions, our data can only cover a part of the variety of music. In our study, we limit ourselves to the English Wikipedia only since it is the largest. Other languages are out of the scope of our work.

Given its status as an online encyclopedia, the content in Wikipedia is dynamic and constantly changing. To get a stable representation we use a database dump of the date 20 July 2023.

3 Experiments

To investigate the potential of the collected free-form text data, we experiment with training a multi-modal deep learning model for text-to-music matching.

Table 3. Top ten most common aspects in the datasets used in this study

Dataset	Top ten aspects
WikiMuTe	classical music, vocals, pop, piano, romantic period classical music, singing, R&B, sings, vocal, piano music
MusicCaps	low quality, instrumental, emotional, noisy, medium tempo, passionate, energetic, amateur recording, live performance, slow tempo

Such a model maps both modalities, text and audio, into the same embedding space to enable comparisons between the modalities. To allow for a quantitative evaluation, we select two downstream tasks: tag-based music retrieval and music auto-tagging. For comparison, we also use the MusicCaps dataset as training data. In the following section, we provide details about the employed model architecture and explain the downstream task evaluation in Sect. 3.2.

3.1   System Overview and Implementation Details

For our system, we adopt a bi-encoder model architecture with an audio tower and a text tower. Each tower consists of an encoder and a shallow multi-layer perceptron (MLP) adapter. The purpose of the MLP adapters is to map the different output dimensionalities of the audio and text encoders to a common embedding space. Figure 3 shows an overview of the system architecture. This architecture allows for joint representation learning of both modalities and easily facilitates cross-modal retrieval since each input modality can be encoded separately. We refer to our system as Music Description Transformer (MDT).

We use the Normalized Temperature-scaled Cross Entropy (NT-Xent) loss [27] – a variant of the InfoNCE loss – to train our bi-encoder model. This loss encourages the embeddings of positive pairs (i.e., matching text and audio samples) to be similar while pushing the embeddings of negative pairs (i.e., non-matching text and audio samples) apart. For a batch of size N, the loss $\mathcal{L}$ is defined as:

$$\mathcal{L}_{NT\text{-}Xent} = \sum_{i=1}^{N} \log \frac{\exp\left(s_{i,i}/\tau\right)}{\sum_{j=1}^{N} \exp\left(s_{i,j}/\tau\right)}, \tag{1}$$

where τ is a temperature hyperparameter, and $s_{i,j}$ is the cosine similarity between the i-th and j-th samples in the batch.

For the encoder of the audio tower, we use the *Music Tagging Transformer* [34]. The authors provide a pretrained model, trained on an autotagging task using the Million Song Dataset (MSD) [4]. The first item of the embedding sequence – referred to as the *CLS* embedding – is used as the encoded audio embedding. The audio encoder expects inputs with a length of 10 s.

For the text encoder, we use a pretrained *SentenceTransformers* language model called `all-MiniLM-L12-v2` [24,32]. This model is produced using knowledge distillation techniques and gives comparable results to other large language

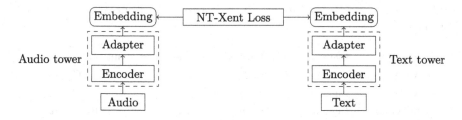

Fig. 3. System overview

models such as BERT [7], but with fewer model parameters and faster inference. To extract the text embeddings, we apply mean pooling on the output sequence.

Both adapters of the two towers employ two-layer MLPs with a layer size of 128 and rectified linear units (ReLU) as activation function after the first layer.

To construct a mini batch of audio-text pairs, we pair audio samples with texts from their associated *aspect* or *sentence* list. All audio clips in MusicCaps are 10 s long, and can directly be passed to the audio tower. For WikiMuTe we randomly select a 10-second section from the audio clip. Texts are formed from the aspect list by joining up to five random aspects, or from the sentence list by selecting one or more sentences. In WikiMuTe, a single sentence is selected from the list at random. Since sentences in MusicCaps often form a coherent paragraph, we select up to n consecutive sentences starting with the first from a sentence list with length S, where $n \in [1 \dots S]$. We use a batch size of 64 and set the temperature parameter τ to 0.07.

We track the mAP@10 metric on the validation set and stop the training if it does not improve for ten epochs. The learning rate is set to a initial value of 0.0001 and divided by a factor of 10 if no improvement was found for five epochs. Finally, after completion of the training, the model weights are reverted to the checkpoint of the epoch with the highest score.

3.2   Experiment Settings and Downstream Task Evaluation

To better compare our dataset to previously published datasets, we include MusicCaps as an additional training dataset. This means we consider two different training datasets: WikiMuTe and MusicCaps. We test three different configurations of WikiMuTe. First, using the text-mined data without any cross-modal relevance filtering, i.e. *no filtering* (WMT_{nf}). Second, a *self-filtered* version where the relevance scoring is done by a model trained with the text-mined data (WMT_{sf}). Third, a version filtered by a system trained on the *MusicCaps* training split (WMT_{mc}).

Text-to-music Retrieval. We want to evaluate our system in a setting that best resembles a real-world music search scenario. Assuming that users would input short queries rather than full sentences into a text-to-music search system, we propose to adopt an aspect-based (or tag-based) retrieval task [35]. This

Table 4. Overview of the datasets used in downstream zero-shot evaluation. The table is adapted from [9].

Dataset	Task	Tracks Count	Tag/Class count	Metric
MTAT$_A$	Tagging	5329	50	ROC/PR
MTAT$_B$	Tagging	4332	50	ROC/PR
MTG-50	Tagging	11 356	50	ROC/PR
MTG-G	Genre	11 479	87	ROC/PR
FMA-Small	Genre	800	8	Acc
GTZAN	Genre	290	10	Acc
MTG-I	Instrument	5115	40	ROC/PR
MTG-MT	Mood/Theme	4231	56	ROC/PR
Emotify	Mood	80	9	Acc

means we use each aspect/tag in the evaluation data as a query and perform retrieval over the audio tracks in the test set. A retrieved track counts as a relevant result if it is labelled with the query.

To evaluate the effectiveness of our approach we utilise the MusicCaps dataset. The official evaluation split contains a subset of 1000 examples that is balanced with respect to musical genres. This balanced subset is set aside as the final test data, while the rest of the evaluation data is used as a validation set for model training. There are 4000 distinct aspect strings in the test dataset. Despite being manually labelled, the dataset contains errors such as typos (e.g., 'fats tempo', 'big bend'). We do not process the text and leave it for future work to clean the data and possibly merge similar labels (e.g., 'no voices', 'no vocals').

To measure the performance on these data, we rely on standard information retrieval metrics computed at different numbers of retrieved documents. Specifically, we use recall at k (R@k) and mean average precision at k (mAP@k) where k denotes the number of results returned for a query, similar to [9,17].

Zero-Shot Music Classification and Auto-Tagging. Additionally, we perform an evaluation in a series of downstream music classification and tagging tasks. This allows us to compare our approach to other systems by using several well-established benchmark datasets. More specifically, we evaluate in a benchmark proposed by [9] using multiple datasets that can be grouped into four tasks: music auto-tagging, genre classification, instrument tagging, and mood labelling. Table 4 lists all datasets used for downstream evaluation and we introduce them briefly.

MTG-Jamendo [5] contains full-length audio tracks and labels taken from an online music platform. The tags are grouped into four sub-categories: *top 50 tags* (MTG-50), *genre* (MTG-G), *instrument* (MTG-I), and *mood/theme* (MTG-MT). We use the test split *split-0*.

MagnaTagATune [15] contains 29-second music clips, each of them annotated with labels collected in a crowdsourcing game. We use the 50 most popular

Table 5. Retrieval scores as percentages on the MusicCaps balanced subset.

Model	Training data	mAP@10	R@01	R@05	R@10
MTR	eCALS	2.8	0.5	2.2	4.3
MDT	WMT_{nf}	3.1 ± 0.3	0.7 ± 0.2	2.7 ± 0.2	4.8 ± 0.5
MDT	WMT_{sf}	3.2 ± 0.2	0.7 ± 0.1	3.0 ± 0.4	5.3 ± 0.4
MDT	WMT_{mc}	3.4 ± 0.2	0.7 ± 0.1	3.1 ± 0.2	5.6 ± 0.4
MDT	MusicCaps	5.0 ± 0.2	1.0 ± 0.1	4.8 ± 0.3	8.7 ± 0.5

tags. Two splits are commonly used [9,16,17]. Similar to [9], we give results for both and refer to them as $MTAT_A$ and $MTAT_B$.

GTZAN [31] is commonly used for genre classification. It contains 30-second music clips labelled with a single genre category. We employ the fault-filtered split of this dataset [29].

Free-Music Archive (FMA) [6] is a large-scale collection of freely available music tracks. We use the *small* subset, which is balanced by genre and includes 30-second high-quality snippets of music with a single genre annotation.

Emotify [2] contains 1-minute music clips from four genres with annotations for induced musical emotion collected through a crowdsourcing game. Each item is annotated with up to three emotional category labels.

Our model is evaluated in a zero-shot setup, which means that it was not optimised for the task. Instead, the classification prediction is proxied by the similarity between an audio and the text representation of a class label or tag.

As metrics we use the receiver operating characteristic area under the curve (ROC-AUC) and the precision-recall area under the curve (PR-AUC) for auto-tagging tasks. These metrics are especially useful in settings where the number of positive examples is much smaller than the number of negative examples. ROC-AUC measures the ability of the model to discriminate between positive and negative examples at different classification thresholds. PR-AUC focuses on the positive class and measures the trade-off between precision and recall at different classification thresholds. For classification tasks, we use accuracy (Acc).

4   Results and Discussion

4.1   Experiment Results

Text-to-music Retrieval. We average results for the text-to-music retrieval evaluation across three randomly initialised training runs and report their mean and standard deviation in Table 5. As a baseline reference, we include results obtained from a pretrained state-of-the-art system (MTR) [9] which was trained in a contrastive learning task on music and tags using BERT as a text encoder.

From the table, it can be seen that all WMT configurations outperform the MTR baseline which scores lowest (mean mAP@10 score of 3.1 to 3.4%

Table 6. Results given as percentages across different downstream tasks

System		Tagging			Instrument
Model	Training data	MTAT-A ROC/PR	MTAT-B ROC/PR	MTG50 ROC/PR	MTG-I ROC/PR
MTR [9]	eCALS	78.4 / 21.2	78.7 / 25.2	76.1 / 23.6	60.6 / 11.3
MDT	WMT$_{nf}$	75.0 / 22.1	75.6 / 26.6	71.5 / 19.7	65.2 / 11.5
MDT	WMT$_{sf}$	75.3 / 23.4	75.8 / 28.0	72.1 / 20.3	66.0 / 12.1
MDT	WMT$_{mc}$	77.9 / 23.5	78.5 / 28.3	73.4 / 21.0	68.8 / 13.7
MDT	MusicCaps	83.1 / 26.9	83.5 / 31.7	74.8 / 20.2	69.6 / 13.7

		Genre			Mood/Theme	
		MTG-G ROC/PR	GTZAN Acc	FMA Acc	MTG-MT ROC/PR	Emot Acc
MTR [9]	eCALS	81.2 / 15.6	87.9	45.1	65.7 / 8.1	33.7
MDT	WMT$_{nf}$	77.9 / 12.5	76.9	37.8	62.9 / 6.4	21.3
MDT	WMT$_{sf}$	78.1 / 12.8	78.2	37.1	63.4 / 6.3	15.0
MDT	WMT$_{mc}$	79.2 / 13.3	75.5	37.1	63.1 / 6.8	15.0
MDT	MusicCaps	75.9 / 11.5	75.9	37.3	65.4 / 7.7	22.5

and 2.8%). The model trained solely on MusicCaps ranks highest. This is not surprising since it is the only one that was trained with data from the same dataset as the test data. Closer inspection of the table shows that relevance filtering consistently improves the results. Both the self-filtered data (WMT$_{sf}$) and the data filtered by MusicCaps (WMT$_{mc}$) improve the scores compared to the system trained on unfiltered data (WMT$_{nf}$).

Zero-Shot Music Classification and Auto-Tagging. We take the best-performing model in each configuration from the previous evaluation and apply it in all zero-shot tasks. Table 6 shows that our method overall achieves competitive results and outperforms the state-of-the-art system (MTR) on some of the benchmark datasets, e.g. for tagging (MTAT-A: PR values of 23.5% and 21.2%) or instrument classification (13.7% and 11.3%). However, the MTR baseline achieves higher scores for the genre and mood/theme tagging tasks as well as MTG50. A possible explanation for this is the much smaller vocabulary in combination with the overall larger dataset size in eCALS (see Table 2). Finally, we notice a similar trend as before: the data cleaned through relevance filtering mostly leads to better results.

4.2 Findings

The most obvious findings to emerge from both evaluations is that the relevance filtering stage is beneficial. This suggests that there is noisy data in the text-mined collection. A manual inspection of the removed texts revealed that it provides a form of semantic filtering that is not possible in the earlier stages of the pipeline. For example, some texts describe parts that are not present in the audio sample, such as the intro or outro or a solo of a song.

Another finding that stands out from the results is that the MusicCaps data generally leads to very good results. The leading scores in the text-to-music retrieval evaluation could be attributed to the fact that the training and test data come from the same data distribution. When comparing the metrics in the zero-shot tasks (an out-of-distribution evaluation) the difference is less evident.

Finally, we find that the WikiMuTe data can be used to achieve competitive results in text-to-music retrieval and some of the tagging classification tasks. These results suggest that datasets containing free-text descriptions can enable complex forms of music analysis and retrieval.

Several factors could explain the observations described above. First, despite the relevance filtering, our data is still noisy and does not match the quality level of hand-labelled data. Second, our data is more sparse, with longer tracks compared to MusicCaps. This disparity results in an imbalance in the ratio of text to audio data. Third, WikiMuTe provides richer data, given that its vocabulary size is double that of MusicCaps. Increasing the amount of audio data would enable us to fully utilise the richness of the text data. Finally, it is also important to note that texts in Wikipedia were not specifically written to provide comprehensive descriptions of the content of music, but rather for a more general purpose. As a result, objective descriptors such as instrumentation are likely better represented than more subjective labels such as moods. In contrast, MusicCaps is a dataset specifically created for music-related content texts. Despite these challenges, we hypothesise that expanding our approach to include more data sources would enhance our results. This assumption is supported by prior studies which demonstrated that larger dataset sizes often lead to superior results, even in the presence of noisy texts [12,16].

5 Conclusion

In this article, we present WikiMuTe, a dataset of rich music descriptions consisting of short labels and long phrases text-mined from Wikipedia articles. We describe how we construct this web-sourced dataset using a three-stage data-mining pipeline. To show the use of the data, we study how it can be leveraged to fine-tune audio and text models to match music audio with musically-relevant textual descriptions. We evaluate our trained system in two ways: tag-based music retrieval and zero-shot music tagging. While the results achieved with a system trained with the WikiMuTe data are generally better or comparable to those of a state-of-the-art system, we observe that data from a manually created dataset can lead to even higher scores. Moreover, we find that filtering the text-mined data according to a cross-modal relevance scoring, leads to improved results. Future work should identify further data sources and apply the proposed techniques to set up a considerably larger dataset.

References

1. Agostinelli, A., Denk, T.I., Borsos, Z., Engel, J.H., et al.: MusicLM: generating music from Text. CoRR abs/2301.11325 (2023)
2. Aljanaki, A., Wiering, F., Veltkamp, R.C.: Studying emotion induced by music through a crowdsourcing game. Inf. Process. Manage. **52**, 115–128 (2016)
3. Bertin-Mahieux, T., Eck, D., Mandel, M.: Automatic tagging of audio: the state-of-the-art. In: Machine Audition: Principles, Algorithms and Systems, pp. 334–352. IGI Global (2011)
4. Bertin-Mahieux, T., Ellis, D., Whitman, B., Lamere, P.: The million song dataset. In: 12th International Society for Music Information Retrieval Conference, ISMIR 2011 (2011)
5. Bogdanov, D., Won, M., Tovstogan, P., Porter, A., Serra, X.: The MTG-Jamendo dataset for automatic music tagging. In: Machine Learning for Music Discovery Workshop, International Conference on Machine Learning (ICML 2019) (2019)
6. Defferrard, M., Benzi, K., Vandergheynst, P., Bresson, X.: FMA: a dataset for music analysis. In: 18th International Society for Music Information Retrieval Conference, ISMIR 2017. pp. 316–323, Suzhou, China (2017)
7. Devlin, J., Chang, M.W., Lee, K., Toutanova, K.: BERT: pre-training of deep bidirectional transformers for language understanding. In: 2019 Conference of the North American Chapter of the Association for Computational Linguistics: Human Language Technologies, Volume 1. Minnesota (2019)
8. Doh, S., Choi, K., Lee, J., Nam, J.: LP-MusicCaps: LLM-based pseudo music captioning. In: 24th International Society for Music Information Retrieval Conference, ISMIR 2023. Milan, Italy (2023)
9. Doh, S., Won, M., Choi, K., Nam, J.: Toward universal text-to-music retrieval. In: 2023 IEEE International Conference on Acoustics, Speech and Signal Processing (ICASSP) (2023)
10. Gemmeke, J.F., Ellis, D., Freedman, D., Jansen, A., et al.: Audio set: an ontology and human-labeled dataset for audio events. In: Proceedings of IEEE ICASSP 2017 (2017)
11. Gruetzemacher, R., Paradice, D.: Deep Transfer learning & beyond: transformer language models in information systems research. ACM Comput. Surv. **54**(10s), 1–35 (2022)
12. Huang, Q., Jansen, A., Lee, J., Ganti, R., et al.: Mulan: a joint embedding of music audio and natural language. In: 23rd International Society for Music Information Retrieval Conference (ISMIR), Bengaluru, India (2022)
13. Huang, Q., Park, D.S., Wang, T., Denk, T.I., et al.: Noise2Music: text-conditioned Music Generation with Diffusion Models. CoRR abs/2302.03917 (2023)
14. Lamere, P.: Social tagging and music information retrieval. J. New Music Res. **37**(2), 101–114 (2008). https://doi.org/10.1080/09298210802479284
15. Law, E., West, K., Mandel, M.I., Bay, M., Downie, J.S.: Evaluation of algorithms using games: the case of music tagging. In: 10th International Society for Music Information Retrieval Conference, ISMIR 2009, Japan (2009)

16. Manco, I., Benetos, E., Quinton, E., Fazekas, G.: Learning music audio representations via weak language supervision. In: 2022 IEEE International Conference on Acoustics, Speech and Signal Processing (ICASSP). IEEE, Singapore (2022)

17. Manco, I., Benetos, E., Quinton, E., Fazekas, G.: Contrastive audio-language learning for music. In: 23rd Internationall Society for Music Information Retrieval Conference (ISMIR), Bengaluru, India (2022)

18. Manco, I., Weck, B., Doh, S., Won, M., Zhang, Y., Bogdanov, D., et al.: The Song Describer Dataset: a Corpus of Audio Captions for Music-and-Language Evaluation. In: Machine Learning for Audio Workshop at NeurIPS 2023 (2023)

19. Manco, I., Weck, B., Tovstogan, P., Bogdanov, D.: Song Describer: a Platform for Collecting Textual Descriptions of Music Recordings. In: Late-Breaking Demo Session of the 23rd Int'l Society for Music Information Retrieval Conf. India (2022)

20. McKee, D., Salamon, J., Sivic, J., Russell, B.: Language-Guided Music Recommendation for Video via Prompt Analogies. In: 2023 IEEE/CVF Conf. on Computer Vision and Pattern Recognition (CVPR). Canada (Jun 2023)

21. Nam, J., Choi, K., Lee, J., Chou, S.Y., Yang, Y.H.: Deep Learning for Audio-Based Music Classification and Tagging: Teaching Computers to Distinguish Rock from Bach. IEEE Signal Process. Mag. **36**(1), 41–51 (2019)

22. Oramas, S., Barbieri, F., Nieto, O., Serra, X.: Multimodal Deep Learning for Music Genre Classification. Transactions of the International Society for Music Information Retrieval **1**, 4–21 (2018). https://doi.org/10.5334/tismir.10

23. Qi, D., Su, L., Song, J., Cui, E., et al.: ImageBERT: Cross-modal Pre-training with Large-scale Weak-supervised Image-Text Data. CoRR abs/2001.07966 (2020)

24. Reimers, N., Gurevych, I.: Sentence-BERT: sentence embeddings using Siamese BERT-networks. In: 2019 Conference on Empirical Methods in Natural Language Processing (2019)

25. Sanh, V., Debut, L., Chaumond, J., Wolf, T.: Distilbert, a distilled version of BERT: smaller, faster, cheaper and lighter. CoRR abs/1910.01108 (2019)

26. Schedl, M., Orio, N., Liem, C.C.S., Peeters, G.: A professionally annotated and enriched multimodal data set on popular music. In: 4th ACM Multimedia Systems Conference, pp. 78–83. ACM, Oslo Norway, February 2013

27. Sohn, K.: improved deep metric learning with multi-class N-pair loss objective. In: Advances in Neural Information Processing Systems, vol. 29. Curran Associates, Inc. (2016)

28. Srinivasan, K., Raman, K., Chen, J., Bendersky, M., Najork, M.: WIT: Wikipedia-based image text dataset for multimodal multilingual machine learning. In: 44th International ACM SIGIR Conference on Research and Development in Information Retrieval. Virtual Event Canada (2021)

29. Sturm, B.L.: The state of the art ten years after a state of the art: future research in music information retrieval. J. New Music Res. **43**(2), 147–172 (2014)

30. Turnbull, D., Barrington, L., Lanckriet, G.: Five approaches to collecting tags for music. In: ISMIR 2008, 9th International Conference on Music Information Retrieval (2008)

31. Tzanetakis, G., Cook, P.: Musical genre classification of audio signals. IEEE Trans. Speech Audio Process. **10**(5), 293–302 (2002)

32. Wang, W., Wei, F., Dong, L., Bao, H., et al.: MiniLM: deep self-attention distillation for task-agnostic compression of pre-trained transformers. In: Advances in Neural Information Processing Systems 33: Annual Conference on Neural Information Processing Systems 2020, NeurIPS 2020. Virtual (2020)

33. Wikipedia contributors: Take On Me - Wikipedia, The Free Encyclopedia (Sep 2023). https://en.wikipedia.org/w/index.php?title=Take_On_Me&oldid=1173253296

34. Won, M., Choi, K., Serra, X.: Semi-supervised music tagging transformer. In: 22nd International Society for Music Information Retrieval Conference, ISMIR 2021 (2021)

35. Won, M., Oramas, S., Nieto, O., Gouyon, F., Serra, X.: Multimodal metric learning for tag-based music retrieval. In: 2021 IEEE International Conference on Acoustics, Speech and Signal Processing (ICASSP), pp. 591–595, June 2021

36. Wu, Y., Chen, K., Zhang, T., Hui, Y., et al.: Large-scale contrastive language-audio pretraining with feature fusion and keyword-to-caption augmentation. In: 2023 IEEE International Conference on Acoustics, Speech and Signal Processing (ICASSP), June 2023

PDTW150K: A Dataset for Patent Drawing Retrieval

Chan-Ming Hsu, Tse-Hung Lin, Yu-Hsien Chen, and Chih-Yi Chiu(✉) ⓘ

Department of Computer Science and Information Engineering, National Chiayi University,
Chiayi, Taiwan
chihyi.chiu@gmail.com

Abstract. We introduce a new large-scale patent dataset termed PDTW150K for patent drawing retrieval. The dataset contains more than 150,000 patents associated with text metadata and over 850,000 patent drawings. We also provide a set of bounding box positions of individual drawing views to support constructing object detection models. We design some experiments to demonstrate the possible ways of using PDTW150K, including image retrieval, cross-modal retrieval, and object detection tasks. PDTW150K is available for download on GitHub [1].

Keywords: Image Retrieval · Cross-Modal Learning · Object Detection

1 Introduction

Conventional patent search systems are mainly based on the text modality. That is, using text query to search text documents. However, a patent document may contain meaningful drawing images conveying important scientific or technical information that is somewhat difficult to be expressed through text. Unlike natural photos, patent drawings are lack of background and contextual information. A set of drawing images constitutes various views of the patent object, providing detailed and high quality representation of the design appearance. We can exploit the rich visual information to provide the image modality in the patent search system for content-based retrieval and cross-modal retrieval.

Most commercial image retrieval systems are mainly designed for natural images. They use image descriptors to find visually-similar images [2–5]. Sketch-based retrieval is a multi-modal problem, where objects are natural images and the query is a sketch represented by the stroke information, to find the association between natural images and sketches [6–9]. In contrast to natural images and sketches, patent drawings receive less attentions; some studies propose hand-crafted descriptors and deep-learning features to represent patent drawings [10–12].

There are a few communities contributing patent drawing benchmarks to advance this research area. MKLab [13] released a dataset of 2,000 patent drawing images extracted from the European Patent Office. These images were manually classified into 41 categories to perform evaluation experiments. Kucer et al. [14] provided a large-scale patent drawing image dataset termed DeepPatent. The dataset consists of over 350,000 public

© The Author(s), under exclusive license to Springer Nature Switzerland AG 2024
S. Rudinac et al. (Eds.): MMM 2024, LNCS 14565, pp. 57–67, 2024.
https://doi.org/10.1007/978-3-031-56435-2_5

domain patent drawings collected from 45,000 design patents published by United States Patent and Trademark Office. These drawing benchmarks are mainly designed for the uni-modal search task, i.e., image-to-image retrieval. They omit the correlation between paired text and drawings in patent documents, which provides an ideal training resource for cross-modal learning. Cross-modal learning [15–18] can supervise the aligned and complementary semantic information in different modalities to learn a better common representation space, and thus can improve the search accuracy for not only uni-modal but cross-modal retrieval. In addition, text-to-image and image-to-text search paradigms provide an alternative search interface for users to find patents between different modalities flexibly. This motivates us to utilize both text and image information to compile a new benchmark of the patent dataset, termed "Patent Drawing: Taiwan Dataset 150K" and abbreviated as "PDTW150K." The dataset contains over 150,000 patents associated with text metadata and more than 850,000 patent drawings. It serves as a large-scale dataset for experimenting both uni-modal and cross-modal retrieval of patent drawings.

Particularly, the patent drawing may contain multiple views showing different appearances of the patent object, such as front view and side view, together with auxiliary information including legends, text segments, and indication lines. Figure 1 gives an example of two drawings of the patent object, where the left one has one view, and the right one has two views. If we can segment multiple views into individual ones and remove noise backgrounds, we can perform a more precise "view-based" matching rather than "drawing-based" matching. However, to manually segment a large amount of views is a labor intensive process. An automatic computer vision technique, such as semantic segmentation or object detection, will be helpful to reduce the human effort. Therefore, we prepare a set of drawings that contains bounding box positions of individual views to support constructing object detection models.

In next sections, we elaborate the PDTW150K dataset and experimental designs to demonstrate possible ways for using it.

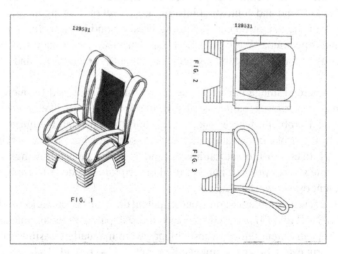

Fig. 1. An example of two patent drawings with auxiliary information, where the left drawing has one view, and the right drawing has two views.

2 Dataset

We introduce the PDTW150K dataset for large-scale patent drawing retrieval experiments. PDTW150K is mainly designed for three purposes, including uni-modal image retrieval (image-to-image), cross-modal retrieval (text-to-image and image-to-text), and object detection. It consists of a total of 153,004 unique patent documents associated with the corresponding metadata description files and 857,757 drawing image files. In addition, 26,748 drawings selected from 4,989 patterns are labeled with bounding box positions of drawing views, which are defined as the foreground category; and the other regions are the background category. The paired text and image set provides different modalities of semantic information, which can be used to learn a common representation space for uni-modal and cross-modal retrieval. The dataset are compiled from the design patent category[1] of Taiwan Patent Database [19] spanning the year from 1990 to 2020, where a design patent is usually illustrated with multiple drawings of various views. The dataset is collected, labeled, and cross-validated by a dozen of undergraduate and graduate students.

2.1 Data Collection

In PDTW150K dataset, each patent consists of the associated metadata and drawings. The metadata include the following fields: patent number (abbreviated as PN), title (TI), issued date (ID), Locarno classification (LC), and abstract (AB). Table 1 gives an example of metadata of the patent "Ice Cube Structure." The Locarno classification defines international classification for industrial designs; it comprises a list of classes and subclasses. The contents of the title and abstract are written in Traditional Chinese. In the dataset, the average length of the abstract is 266 characters, and the average number of drawings is 5.6 per patent document. Figures 2 and 3 show the statistic information for the patent distribution per LOC class and per year, respectively. The class distribution is highly imbalanced, making the classification task more challenging.

We partition PDTW150K dataset into large and small parts without overlap. Each part is divided into train, validation, and test sets with class stratification. The numbers of patents and drawings of these sets are listed in Table 2. In each sets, we pack patents into folders according to their patent numbers; each folder contains the corresponding patent metadata and drawings. Partial patent folders contain additional bounding box positions of drawing views. For the retrieval task, the test set is set aside to serve as a patent database to search through. To perform image-to-image and image-to-text search, the first drawing of each folder in the test set is designated as the query image.

---

[1] The other categories are invention pattern and utility model pattern.

2.2 Dataset Availability and Distribution

The PDTW150K dataset is available for download on Github [1]. The dataset includes all drawing images in PNG format and metadata in CSV format. PDTW150K is provided under "Open Government Data License, version 1.0 (OGDL-Taiwan-1.0)" in a free of charge, non-exclusive, and sublicensable method for the public [20].

Table 1. An example of metadata of the patent "Ice Cube Structure."

Patent Number (PN)	TW127824
Title (TI)	冰塊結構 (Ice Cube Structure)
Issued Date (ID)	19900121
Locarno classification (LC)	01-01
Abstract (AB)	圖1為本新式樣之「冰塊結構」之形狀;圖2為其正面圖;圖3為其頂面圖;圖4為其側面圖,另一側面相同;圖5為其底面圖。本創作之特徵在於其整體之造型,其中:一較大角錐體中具有四個較小之角錐體,該諸較小角錐體間以形成於其間之錐谷子以分隔。大角錐體之外緣係為圓弧狀。整體觀之,本創作確為一特異之作。

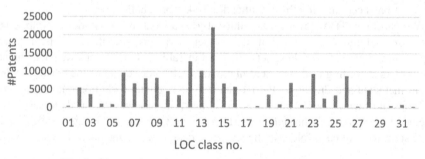

Fig. 2. The number of patents per LOC class (totally 32 classes).

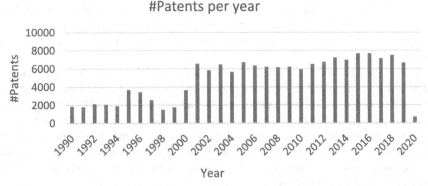

Fig. 3. The number of patents per year (from 1990 to 2020).

Table 2. The numbers of patents and drawings of the train, validation, and test sets in large and small parts of PDTW150K.

#patents	Large	Small	#drawings	Large	Small
Train	79,399	14,979	Train	448,316	80,330
Validation	19,958	4,998	Validation	113,284	26,586
Test	27,915	5,000	Test	157,838	26,947
Total	127,272	24,977	Total	719,438	133,863

3 Experiments

We design the following three experimental tasks based on PDTW150K.

- Image retrieval: performing image-to-image retrieval that finds similar drawings for the given query drawing.
- Cross-modal retrieval: performing text-to-image and image-to-text retrieval that finds similar drawings (text) for the given query text (drawing).
- Object detection: performing object detection in a drawing image that predicts bounding box positions of drawing views.

All experiments were performed on the small part dataset. The experimental setup and results are described in the following.

3.1 Image Retrieval

Figure 4 shows our flow of image retrieval. The pipeline is to transform the query drawing to an embedding through the proposed model, which is a CNN-based model (VGG16 [21] and ResNet50 [22] are used in this case) followed by the NetVLAD layer [4] and under the supervision of triplet loss training. The number of embedding dimensions is 2048. The training parameters are set as follows: the optimizer is Adam; the triplet loss margin is 0.28; the learning rate is 1×10^{-6}; the batch size is 32; and the number of epochs is 100. The Euclidean distances between the query and database embeddings are computed to return the top K drawings with the smallest distances. If the retrieved drawings have the same LOC class and subclass as the query, they are considered true positives.

The retrieval accuracy is evaluated by the metric mAP@K (mean Average Precision at K), which is calculated in the same way as Product1M [23]. The result is listed in Table 3. We find that ResNet50 yields a better accuracy than VGG16. Concatenating NetVLAD in VGG16 can improve the accuracy, but it does not work well in ResNet50. It might incur overfitting in the more complicated model. Figure 5 gives two examples of patent drawing retrieval. In each row, the leftmost image is the query drawing, and the remaining images are the retrieved drawings. On the top of each image is labeled with its LOC class and subclass.

Table 3. The image retrieval accuracy.

mAP	VGG16	VGG16 + NetVLAD	ResNet50	ResNet50 + NetVLAD
@1	0.654	0.670	0.707	0.682
@5	0.670	0.687	0.721	0.697
@10	0.639	0.655	0.687	0.663
@20	0.589	0.604	0.638	0.611
@30	0.552	0.569	0.602	0.574
@40	0.524	0.543	0.576	0.547
@50	0.503	0.521	0.555	0.525

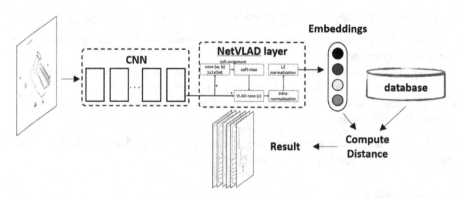

Fig. 4. The flow of image retrieval.

3.2 Cross-Modal Retrieval

We construct a cross-modal representation model, as illustrated in Fig. 6. For the text modality, the metadata of patent title and excerpted abstract (the first 100 characters)

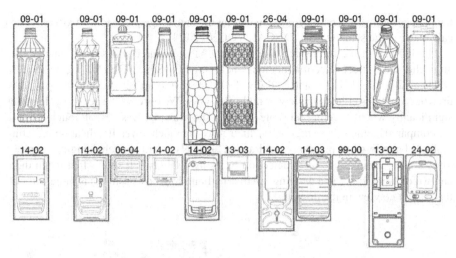

Fig. 5. Examples of the image retrieval result.

are used to fine-tune the pre-trained Chinese BERT model. For the image modality, we employ ResNet50 to learn the drawing representation space. The number of dimensions of the common representation space is 768. The total loss combines MSE loss, triplet loss, and classification loss expressed by:

$$L_{total} = L_{MSE} + \lambda_1 \cdot L_{triplet} + \lambda_2 \cdot L_{class},$$

where λ_1 and λ_2 are weights. The MSE loss L_{MSE} calculates the Euclidean distance between the paired text and image of the same patent. The triplet loss $L_{triplet}$ computes the relative distances between anchor, positive and negative examples, where the example is defined as positive if its LOC class is the same as the anchor; otherwise it is negative. The classification loss L_{class} is actually the cross entropy between the class predictions of the image modality. The training parameters are set as follows: the optimizer is Adam; the triplet loss margin is 0.1; the learning rate is 1×10^{-5}, the batch size is 16; the number of epochs is 100; and $\lambda_1 = 1$ and $\lambda_2 = 0.02$.

Table 4 lists the accuracy of cross-modal retrieval in terms of image-to-text and text-to-image. Recall@K is used as the accuracy metric, which is calculated in the same way as Recipe1M [15]. Three different combinations of the loss functions are evaluated as ablation study. The result indicates the combination of the three losses can raise the recall rates effectively.

3.3 Object Detection

26,748 drawings are labeled with the bounding box positions of drawing views. They are split into the training set, validation set, and test set at ratio 7:1:2. Hence the training set contains 18,760 drawings, the validation set contains 2,658 drawings, and the test set contains 5,330 drawings. Note that we define only two categories in the object detection task: the bounding box region is the foreground, whereas the other region is the background. The accuracy is evaluated by mean average precision (mAP), which is calculated in the same way as PASCAL VOC2012 [24].

Table 5 lists the performance under different APs compared with several state-of-the-art anchor-free object detection methods, including FCOS [25], DETR [26], ATSS [27], DyHead [28], MFFPN [29], and YOLOv7 [30]. The results are divided into two parts: the drawings contains either one view or multiple views. We observe the accuracy degrades significantly when detecting multiple views in the drawing. Besides, although most of the compared methods have high AP_{50}, their AP_{90} are much lower. It indicates predicting multiple bounding box positions in a drawing precisely is still very challenging. Figure 7 gives some examples of object detection by using YOLOv7. It shows that when there are two or more views in a drawing, predicting bounding box positions is more difficult than only one view in the drawing.

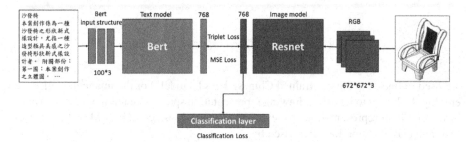

Fig. 6. The proposed cross-modal representation model.

Table 4. The cross-modal retrieval accuracy.

Image-to-Text			
Recall	L_{MSE}	$L_{MSE} + L_{triplet}$	$L_{MSE} + L_{triplet} + L_{class}$
@1	0.058	0.163	0.180
@5	0.193	0.327	0.341
@10	0.424	0.440	0.428
Text-to-Image			
Recall	L_{MSE}	$L_{MSE} + L_{triplet}$	$L_{MSE} + L_{triplet} + L_{class}$
@1	0.035	0.102	0.149
@5	0.267	0.310	0.350
@10	0.435	0.440	0.485

Table 5. The object detection accuracy.

The drawing contains only one view (4,001 test drawings)

mAP	FCOS	DETR	ATSS	DyHead	MFFPN	YOLOv7
AP_{50}	0.982	0.955	0.977	0.979	0.981	0.980
AP_{75}	0.890	0.724	0.883	0.906	0.921	0.916
AP_{90}	0.378	0.143	0.252	0.358	0.551	0.564

The drawing contains multiple views (1,329 test drawings)

mAP	FCOS	DETR	ATSS	DyHead	MFFPN	YOLOv7
AP_{50}	0.962	0.105	0.966	0.830	0.968	0.958
AP_{75}	0.648	0.005	0.731	0.637	0.762	0.768
AP_{90}	0.043	0.000	0.086	0.093	0.139	0.151

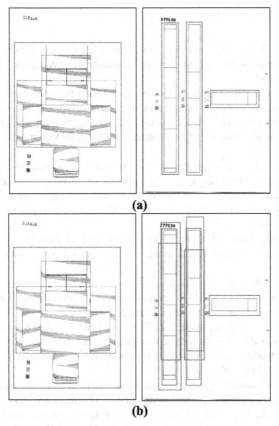

Fig. 7. Examples of the object detection result. (a) Ground truth of view bounding boxes. The left drawing contains one view, and the right drawing contains three views; (b) the bounding boxes predicted by using YOLOv7.

4 Conclusions

In this paper, we introduce a new released patent drawing dataset PDTW150K, which contains more than 150,000 patents associated with text metadata and over 850,000 patent drawings. Besides, partial drawings provide the bounding box labels of drawing views. We design some experiments to demonstrate the ways of using PDTW150K, including image retrieval, cross-modal retrieval, and object detection. The results demonstrate PDTW150K is a valuable resource to push the advancement in patent drawing retrieval. Possible future directions may be adding more patent datasets from other regions/countries, and providing more precise label information for the semantic segmentation of drawing views.

Acknowledgments. This work was supported by the National Science and Technology Council under grants NSTC 112-2221-E-415-008-MY3.

References

1. Patent Drawing: Taiwan Dataset 150K. https://github.com/ncyuMARSLab/PDTW150K
2. Lowe, D.G.: Distinctive image features from scale-invariant keypoints. Int. J. Comput. Vision **60**(2), 91–110 (2004)
3. Dalal, N., Triggs, B.: Histograms of oriented gradients for human detection. In: Proceedings of IEEE Computer Society Conference on Computer Vision and Pattern Recognition (2005)
4. Arandjelović, R., Zisserman, A.: NetVLAD: CNN architecture for weakly supervised place recognition. In: Proceedings of IEEE Conference on Computer Vision and Pattern Recognition (2016)
5. Dosovitskiy, A., et al.: An image is worth 16×16 words: transformers for image recognition at scale. In: Proceedings of International Conference on Learning Representations (2021)
6. Zhang, H., Liu, S., Zhang, C., Ren, W., Wang, R., Cao, X.: Sketchnet: sketch classification with web images. In: Proceedings of IEEE Conference on Computer Vision and Pattern Recognition (2016)
7. Bui, T., Ribeiro, L., Ponti, M., Collomosse, J.: Sketching out the details: sketch-based image retrieval using convolutional neural networks with multi-stage regression. Comput. Graph. **71**, 77–87 (2018)
8. Chowdhury, P.N., Bhunia, A.K., Gajjala, V.R., Sain, A., Xiang, T., Song, Y.Z.: Partially does it: towards scene-level FG-SBIR with partial input, In: Proceedings of IEEE Conference on Computer Vision and Pattern Recognition (2022)
9. Chaudhuri, A., Bhunia, A.K., Song, Y.Z., Dutta, A.: Data-free sketch-based image retrieval. In: Proceedings of IEEE Conference on Computer Vision and Pattern Recognition (2023)
10. Vrochidis, S., Moumtzidou, A., Kompatsiaris, I.: Concept-based patent image retrieval. World Patent Inf. **34**(4), 292–303 (2012)
11. Ni, H., Guo, Z., Huang, B.: Binary patent image retrieval using the hierarchical oriented gradient histogram. In Proceedings of International Conference on Service Science (2015)
12. Jiang, S., Luo, J., Pava, G.R., Hu, J., Magee, C.L.: A convolutional neural network-based patent image retrieval method for design ideation. In: Proceedings of International Design Engineering Technical Conferences and Computers and Information in Engineering Conference 83983 (2020)
13. Patent Images Databases. https://mklab.iti.gr/results/patent-image-databases/

14. Kucer, M., Oyen, D., Castorena, J., Wu, J.: DeepPatent: large scale patent drawing recognition and retrieval. In: Proceedings of IEEE Winter Conference on Applications of Computer Vision, pp. 2309–2318 (2022)

15. Marın, J., et al.: Recipe1m+: a dataset for learning cross-modal embeddings for cooking recipes and food images. IEEE Trans. Pattern Anal. Mach. Intell. **43**(1), 187–203 (2021)

16. Tu, R.C., Mao, X.L., Ji, W., Wei, W., Huang, H.: Data-aware proxy hashing for cross-modal retrieval. In: Proceedings of ACM International Conference on Information Retrieval, pp. 686–696 (2023)

17. Wu, X., Lau, K., Ferroni, F., Osep A., Ramanan, D.: Pix2Map: cross-modal retrieval for inferring street maps from images. In: Proceedings of IEEE International Conference on Computer Vision and Pattern Recognition, pp. 17514–17523 (2023)

18. Xu, P., Zhu, X., Clifton, D.: Multimodal learning with transformers: a survey. IEEE Trans. Pattern Anal. Mach. Intell. **45**(10), 12113–12132 (2023)

19. Taiwan Patent Search System. https://twpat.tipo.gov.tw/

20. Open Government Declaration. https://www.tipo.gov.tw/en/cp-389-800409-65ea6-2.html

21. Simonyan, K., Zisserman, A.: Very deep convolutional networks for large-scale image recognition. In: Proceedings of International Conference on Learning Representations (2015)

22. He, K., Zhang, X., Ren, S., Sun, J.: Deep residual learning for image recognition. In: Proceedings of IEEE Conference on Computer Vision and Pattern Recognition (2016)

23. Zhan, X., et al.: Product1m: towards weakly supervised instance-level product retrieval via cross-modal pretraining. In: Proceedings of IEEE International Conference on Computer Vision (2021)

24. Everingham, M., Van Gool, L., Williams, C. K. I., Winn, J., Zisserman, A.: The PASCAL Visual Object Classes Challenge 2012 (VOC2012) Results. http://host.robots.ox.ac.uk/pascal/VOC/voc2012/

25. Tian, Z., Shen, C., Chen, H. He, T.: FCOS: fully convolutional one-stage object detection. In: Proceedings of IEEE International Conference on Computer Vision (2019)

26. Carion, N., Massa, F., Synnaeve, G., Usunier, N., Kirillov, A., Zagoruyko, S.: End-to-end object detection with transformers. In: Vedaldi, A., Bischof, H., Brox, T., Frahm, J.-M. (eds.) Computer Vision – ECCV 2020. LNCS, vol. 12346, pp. 213–229. Springer, Cham (2020). https://doi.org/10.1007/978-3-030-58452-8_13

27. Zhang, S., Chi, C., Yao, Y., Lei, Z., Li, S.Z.: Bridging the gap between anchor-based and anchor-free detection via adaptive training sample selection. In: Proceedings of IEEE Conference on Computer Vision and Pattern Recognition (2020)

28. Dai, X., et al.: Dynamic head: unifying object detection heads with attentions. In: Proceedings of IEEE Conference on Computer Vision and Pattern Recognition (2021)

29. Chen, Y.H., Chiu, C.Y.: MFFPN: an anchor-free method for patent drawing object detection. In: Proceedings of International Conference on Machine Vision Applications (2023)

30. Wang, C.Y., Bochkovskiy, A., Liao, H.Y.M.: YOLOV7: trainable bag-of-freebies sets new state-of-the-art for real-time object detectors. In: Proceedings of IEEE Conference on Computer Vision and Pattern Recognition (2023)

Interactive Question Answering
for Multimodal Lifelog Retrieval

Ly-Duyen Tran[1][✉], Liting Zhou[1], Binh Nguyen[2,3], and Cathal Gurrin[1]

[1] Dublin City University, Dublin, Ireland
allie.tran@adaptcentre.ie
[2] AISIA Research Lab, Ho Chi Minh, Vietnam
[3] Ho Chi Minh University of Science, Vietnam National University, Hanoi, Vietnam

Abstract. Supporting Question Answering (QA) tasks is the next step for lifelog retrieval systems, similar to the progression of the parent field of information retrieval. In this paper, we propose a new pipeline to tackle the QA task in the context of lifelogging, which is based on the open-domain QA pipeline. We incorporate this pipeline into a multimodal lifelog retrieval system, which allows users to submit questions prevalent to a lifelog and then suggests possible text answers based on multimodal data. A test collection is developed to facilitate the user study, the aim of which is to evaluate the effectiveness of the proposed system compared to a conventional lifelog retrieval system. The results show that the proposed system is more effective than the conventional system, in terms of both effectiveness and user satisfaction. The results also suggest that the proposed system is more valuable for novice users, while both systems are equally effective for experienced users.

Keywords: Lifelogging · Question answering · Human-Computer Interaction

1 Introduction

Lifelogging has gained significant attention in recent years as a means of digitally recording and preserving one's life experiences. Lifelog data typically consists of multimodal information, including text, images, audio, and video, and presents unique challenges for efficient retrieval and organisation due to the volumes of multimodal data collected. To address these issues, pioneering research challenges and competitions have been organised, such as Lifelog Search Challenges [26] and NTCIR Lifelog tasks [34]. The Lifelog Search Challenge (LSC) is one of the most well-known benchmarking activities in the field, which has been running annually since 2018, with the aim of advancing the state-of-the-art in lifelog retrieval systems in an open, metrics-based manner. The challenge focuses on comparing novel techniques for supporting users to efficiently search for a specific moment in their lifelogs. However, most research in the area has heretofore focused on interactive retrieval systems, while the Question Answering

© The Author(s), under exclusive license to Springer Nature Switzerland AG 2024
S. Rudinac et al. (Eds.): MMM 2024, LNCS 14565, pp. 68–81, 2024.
https://doi.org/10.1007/978-3-031-56435-2_6

(QA) challenge remains an under-explored topic. With the increasing prevalence of pervasive computing, there is a need to integrate question answering (QA) capabilities into lifelog retrieval systems, allowing users to ask specific questions about their lifelogs and receive text-based or spoken answers. Understanding this need, a QA task was introduced at LSC'22 [9], accepting images as the answers, and thus it was not a true QA task. In LSC'23 [10], the QA task was fully integrated, meaning that QA topics were answered by text-based submissions. This highlights the direction of the research community, which is moving towards supporting lifelog QA systems.

Although a lifelog QA dataset, LLQA [24], was constructed in order to gain more attention for the task, all questions in this dataset are pertained to short, provided snippets of lifelog data. In other words, the questions are related to a particular activity at a point in time, for example, 'What is the lifelogger holding in his hand [at this particular time]?'. In reality, more open-ended type of questions that span any period of time, from a moment to a lifetime, are more prevalent, such as 'What is my favourite drink?' or 'Where did I go on holiday last summer?'. Therefore, a comprehensive question dataset is therefore necessary to address this issue. In this paper, we present a test collection for the QA task, utilising all the published datasets in the LSC spanning from 2016 to 2023. By leveraging the existing LSC datasets, we aim to create a comprehensive test collection that addresses the broader range of lifelog queries users may have. Moreover, we propose a pipeline to incorporate QA capabilities into an existing lifelog retrieval system. We conduct a user study to evaluate the effectiveness and user satisfaction of our proposed lifelog QA system and compare it to a baseline search-only approach. Our preliminary results demonstrate the superiority of our proposed system over the baseline approach, with significant improvements in both effectiveness and user satisfaction metrics. Furthermore, the results suggest that our proposed system is particularly well-suited for novice users, offering a more intuitive and efficient lifelog retrieval experience. The following sections of this paper will describe the details of our proposed pipeline, the construction of the test collection, the user study methodology, and the comprehensive analysis of the results obtained.

The contributions of this paper are thus as follows: (1) a novel pipeline for multimodal lifelog QA systems, drawing inspiration from the open-domain QA pipeline; (2) a test collection for the lifelog QA task comprising 235 questions sourced from the LSC datasets; and (3) a user study assessing the effectiveness and user satisfaction of our lifelog QA system in comparison to a search-only baseline approach.

2   Related Work

2.1   Lifelog Retrieval

Retrieval systems for lifelog data have been a popular research topic for almost a decade now, since the first NTCIR-lifelog challenge in 2015 [8]. It is a crucial task in order to manage and make use of the large amount of multimodal data collected by lifeloggers. The seminal lifelog retrieval system is MyLifeBits [7], which supported limited full-text search, text and audio annotations, and hyperlinks. Since then, many other lifelog retrieval systems have been proposed and evaluated. The Lifelog Search Challenge (LSC) is an annual benchmarking campaign that aims to advance the state-of-the-art in lifelog retrieval systems. The dominant approach of the participating teams has been focusing on concept-based techniques, leveraging computer vision models to automatically extract visual analysis from lifelog images, such as object recognition, scene understanding, and optical character recognition (OCR). The outputs of these models, also known as 'concepts', are then used in accompanying metadata (e.g. timestamps, GPS coordinates, etc.) for indexing and retrieval. Various ranking techniques borrowed from the field of text-based information retrieval have been explored, such as TF-IDF [28], BM25 [3,29], bag-of-words (BoW) [19] to rank the lifelog moments based on the concepts. Other metadata such as timestamps and location information are also used to improve the retrieval performance by boolean filtering [23] or map visualisation [28]. Recently, with the rise of cross-modal embedding models, such as CLIP [20] and CoCa [33], large-scale pretrained models have been utilised to extract the visual and textual features from image contents and questions, and then provide a similarity score between the features to rank the lifelog moments. This embedding-based approach allows a more user-friendly experience by allowing users to search for lifelog data using natural language queries and significantly improves the retrieval performance [1,27]. As a result, most conventional search tasks in the LSC are considered mostly solved by this embedding-based approach. This allowed the organisers to introduce the lifelog QA task in the LSC'22, aiming to evaluate the effectiveness of lifelog retrieval systems in answering questions about lifelog data. Since QA is a relatively new task in the lifelogging domain, there is a lack of research in this area. Our study aims to contribute to this area by proposing a pipeline for integrating QA capabilities into lifelog systems and evaluating its effectiveness compared to a baseline search-only approach.

2.2 Open-Domain Question Answering

Open-domain QA (OpenQA) is the task of answering questions without any specified context, as opposed to machine reading comprehension (MRC) where specified context passages are provided. Most modern OpenQA systems follow a 'Retriever-Reader' architecture [4,35] which contains a *Retriever* and a *Reader*. Given a question, the *Retriever* is responsible for retrieving relevant documents to the question in an open-domain dataset such as Wikipedia and the World Wide Web (WWW); while the *Reader* aims at inferring the final answer from the received documents, which is usually a neural MRC model. Specifically, the *Retriever* can utilise traditional information retrieval techniques such as TF-IDF [4] and BM25 [32], or more advanced deep retrieval models to encode the question and documents [12,15]. After that, approaches for the *Reader* can be categorised into extractive and generative models. Extractive models [4,12,32] are designed to extract an answer span from the retrieved documents using

BERT [5], RoBERTA [18], etc. On the other hand, generative approaches [11,17] apply models such as BART [16] and T5 [21] to generate the answer in an open-ended manner. To further extend the architecture, some works [14,30] have proposed to re-rank the retrieved documents before feeding them into the *Reader* [14], or train the entire OpenQA system in an end-to-end manner [15,17].

In this paper, we take inspiration from open-domain QA research to design a lifelog QA system for the following reasons: (1) the lifelog QA task is similar to the open-domain QA task in the way that the questions are not limited to a specific event or image, but the whole lifelog; and (2) the two-stage architecture is flexible enough to incorporate with existing state-of-the-art lifelog retrieval systems without the need to re-train the entire system.

3   QA Test Collection

In order to compile a comprehensive QA test collection, we utilise the largest two lifelog datasets in the LSC, namely LSC'21 [26] and LSC'22 [9]. Together, these datasets feature an extensive repository of lifelogging data collected by one lifelogger. This data encompasses various types of multimodal information, including over 900,000 point-of-view images, music listening history, biometrics, and GPS coordinates.

As the time of writing, there are 19 official QA information needs (topics) posed by the lifelogger who created the datasets for the LSC challenge (8 in LSC'22 and 11 in LSC'23). In addition to these, we have created a larger collection of topics to include more variety in the user study, leading to 235 questions in total. These questions were inspired by the official known-item search (KIS) topics in all LSCs from 2019 to 2023. An example KIS topic is '*I was building a computer alone in the early morning on a Friday at a desk with a blue background. Sometimes I needed to refer to the manual. I remember some Chinese posters on the desk background. I was in Dublin City University in 2015*'. For each topic, we identified the relevant lifelog data that were provided by the organisers, including time, location, and lifelog images. We then created questions based on the information in the topic description and the provided data. For example, one question for the above topic is '*How many days did it take for me to build my computer back in March 2015?*', whose answer, '*2 d*', can be found by looking at the timestamps of the ground-truth images. After that, each question in the collection is labelled based on the type of information that is asked, such as Location, Time, and Colour. The test collection focuses on questions that have specific answers, which are either a single word or a short phrase, with as little ambiguity as possible. This is to ensure that the answers can be easily evaluated. The questions are also designed to be as diverse as possible, to cover different types of information that can be retrieved from a lifelog. Thus, we propose 8 different types of questions for this collection as follows:

– **Location**: these are questions that ask about the name of a country, a city, or a venue (e.g. restaurant) where some specific events happened. For example, 'Where did I go the get my car repaired in 2020?';

- **Object**: the answers generally refer to some objects that are involved in the events. For example, 'What did I eat for dinner on the 1st of January 2020?'
- **Counting**: these require counting the number of people or things that appeared in an event. For example, 'How many different papers did I read on the plane going to Germany back in June?'
- **Time**: these are questions that ask about the date/time of some events. For example, 'When did I last go to the zoo?' or 'What time did I go shopping for emergency supplies in 2020?'
- **Frequency**: these require counting the number of times some activities happened. For example, 'How many times did I have BBQs in my garden in the summer of 2015?'
- **OCR**: the answers are some texts that appeared in the lifelog images. For example, 'Which airline did I fly with most often in 2019?' requires reading the boarding passes or the airlines' brochures on the back of the seats.
- **Colour**: these are questions that ask about the colour of some objects. For example, 'What colour was the rental car I drove before 2018?'
- **Duration**: the answers are the duration of some events. For example, 'How long did it take me to drive from Dublin to Sligo in 2016?'

The distribution of the questions in the collection is shown in Fig. 1. Time and Location are the most common types of questions, which is to be expected. The least common type is Frequency, which possibly is because it is difficult to verify the answer in a short time, which is not suitable for the user study. The full list of questions and their answers is available at https://docs.google.com/spreadsheets/d/1eTlKfurPg0LOT-PDkf3SpctdkvrlyV_u1v3IOdgU4wU.

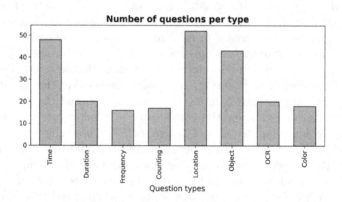

Fig. 1. Distribution of the questions in the test collection.

4   Lifelog Question Answering Pipeline

Inspired by the open-domain QA pipeline [4], we formally propose a pipeline for the lifelog QA system as shown in Fig. 2. Two key components of the pipeline are

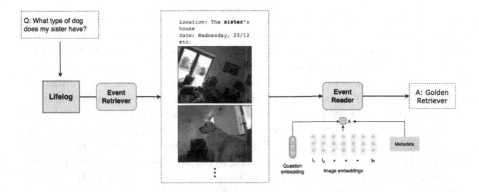

Fig. 2. The proposed pipeline for the QA system. The Event Retriever is in charge of retrieving the lifelog data that are relevant to the given question. On the other hand, the Event Reader component is responsible for generating answers based on the retrieved data.

(1) Event Retriever and (2) Event Reader. The Event Retriever is in charge of retrieving the lifelog data that are relevant to the given question. On the other hand, the Event Reader component is responsible for generating answers based on the retrieved data. This pipeline is designed to be flexible so that different retrieval and QA methods can be used. As a result, it can seamlessly integrate with most existing lifelog retrieval systems, serving as the initial component in the process.

4.1 Event Retriever

The first component is a crucial part of the pipeline, as it is responsible for retrieving the relevant lifelog information that is used to generate the answers. Given a question, the lifelog retrieval component determines the relevance of events in the lifelog to the question based on various multimodal features, such as time, location, and image content. The events are then ranked using a suitable ranking method as seen in a conventional lifelog retrieval system, namely boolean filtering, text-based retrieval, or embedding-based retrieval as described in Sect. 2.1.

To adapt conventional image-focus lifelog retrieval systems, a post-processing step might be useful to aggregate the information from the retrieved data and reduce the amount of information to be passed to the question answering component, which is important for the efficiency of the system. Grouping data that belong to the same event is a possible approach, which can be done by clustering the retrieved events based on their time and location information.

Our proposed system is built upon MyEachtra[25], which participated in LSC'23 and achieved the second-best overall performance. Location and time information are extracted directly from the question and used to filter the events. The remaining part of the question is encoded by the text encoder from OpenAI CLIP [20] and is used to rank the events based on similarity scores. The main difference between MyEachtra from other conventional lifelog retrieval systems is that it expands the unit of retrieval from point-in-time moments to a longer period of time, or 'events', aiming to reduce the search space and provide more lifelog context to the user. This also allows the system to support more complex queries, such as questions about duration and frequency, which are difficult to answer without any organisation of the lifelog data. Since MyEachtra is event-focus, the post-processing step described above is not necessary.

The top-ranked events are then passed to the question answering component to generate the answers. The cut-off point for the number of events to be passed to the question answering component is a hyperparameter of the system, which can be tuned to achieve the best performance. It is also important to note that different types of questions may require different numbers of events to be passed to the question answering component. For example, questions that require counting the Frequency of some events may require more events to be passed to the question answering component than questions that ask about the Location of some events. In this paper, we use the top 10 events as the default cut-off point for all types of questions to simplify the process. However, this can be adjusted in the future to improve the performance of the system.

4.2   Event Reader

This QA component of the pipeline is responsible for generating the answers based on the retrieved events. The answers are generated by combining the information from the retrieved events and the question. The information from the retrieved events can be extracted from the metadata, such as time and location, or the image content, such as OCR text. To address the multimodality of the lifelog data, we propose an ensemble of two different models to handle both visual and non-visual information. The original MyEachtra system proposed using video QA models and treating each event as a video clip with a very low frame rate. This allows the system to leverage both the visual content and the temporal relationship between the images in the events. However, this model is not suitable for questions that do not require visual information, such as questions about Time and Location. To address this issue, we propose to add a text-only QA model to handle non-visual information. Finally, the two models are combined to generate the suggested answers which are shown to the user.

Specifically, in this paper, FrozenBiLM [31] is employed as the VideoQA model, which builds on a frozen bidirectional language model as well as a frozen visual encoder, CLIP [20]. FrozenBiLM was pretrained on a large-scale video-caption pairs dataset WebVid10M [2]. As it builds on a language model, Frozen-BiLM can be used to predict the most probable answer given the question as a masked prompt, such as '[CLS] Question: <Question>? Answer: [MASK]'. We also experimented on finetuning FrozenBiLM on the LLQA dataset [24], however, the performance does not improve due to the small size of the dataset. Thus, we use the model that was fine-tuned on the ActivityNet-QA dataset [6] instead.

The new addition to the model is the use of the text-only QA model to handle non-visual information. Related information from the metadata is used to generate a contextual paragraph in the format of 'The event happened at <location> on <date>, starting at <time> and ending at <time>. Text that can be read from the images includes: <OCR text>'. We use the RoBERTA model [18], pretrained on SQuAD 2.0 [22] to predict the answer span from the generated paragraph.

5   User Study

To evaluate the effectiveness of the proposed lifelog QA system, we conducted a user study comparing the performance of the QA system to a baseline search-only system. This allows for a direct comparison between the two systems, providing insights into the effectiveness of the QA system and the potential to improve the lifelog retrieval experience.

5.1   Setup

A total of 10 participants, with ages ranging from 20 to 35, were recruited for the user study. All participants have basic computer skills, with very little familiarity with the concept of lifelog retrieval and question answering. The participants were randomly to one of the two groups: the baseline group and the QA group. The baseline group was asked to use the baseline system first, then the QA system, and vice versa for the QA group. This is to ensure that the order of the systems does not affect the results.

Each participant had a training period of 10-15 min to get familiar with the concept of lifelogging and the systems before the test. For each system, the participants were asked to use the system to answer 8 randomly selected questions from the test collection, one for each type of question. Three minutes were given for each question. If the participants were sure about the answer, they could submit it and the judging system (controlled by a real-time human judge) would inform them whether the answer was correct or not. If the answer was incorrect, the participants were asked to try again. If they could not find the answer within 3 min, they were asked to move on to the next question. The participants were also asked to fill in a questionnaire after using each system. The questionnaire is based on the User Experience Questionnaire (UEQ) [13], which is a standard questionnaire for evaluating the usability of a system. The questionnaire consists of 8 questions, each of which is rated on a scale of -3 to 3 (with 0 as the neutral score). Feedback on the system was also encouraged, which is used to improve the system in the future.

Similarly to the LSC, the performance of the systems is measured based on (1) the accuracy of the answers, (2) the number of wrong submissions, w, and (3) the time taken to answer the questions, t. For each task, if it is solved (the correct answer was submitted), the score s is calculated as follows:

$$s = 100 - 50 \times \frac{t}{180} - 10 \times w \tag{1}$$

If the task is not solved, $s = 0$.

5.2 Baseline System

The baseline system used in this user study is a lifelog retrieval system that is also the LSC'23 baseline system, E-Myscéal [27], an embedding-based variation of the original Myscéal [28]. Myscéal and its upgraded versions have been the best-performing system in the LSC since 2020 and participated in LSC'23 as a baseline system for benchmarking other lifelog systems. It is designed to accommodate novice users by accepting full sentences as search queries. A query parsing component is used to extract the relevant information from the query, such as location, time, and visual information. The extracted information is then used to compose Elasticsearch queries to retrieve the relevant images. The retrieved images are then ranked based on their relevance to the query. The mechanism to retrieve the textual data field is BM25, while the mechanism to retrieve the visual data field is the cosine similarity between the query and the image features. The query and image features are extracted using the OpenAI CLIP model [20].

The user can also browse the lifelog images using a popover timeline, which is shown when the user clicks on any image shown on the result page. The popover timeline shows the images taken before and after the selected image, which allows the user to browse the images in chronological order. The user can also click on any image in the popover timeline to view the image in full size. More features to support the user in the lifelog retrieval task are also provided, such as the ability to search for visually similar images, filter the results by map location, and most importantly, search for temporally related queries.

5.3 QA System

We use the proposed pipeline to integrate QA capabilities into the Myscéal system by (1) shifting the unit of retrieval to events, which is the main difference between MyEachtra and E-Myscéal [28] in the retrieval stage; and (2) adding a QA component to generate the answers based on the retrieved events. Refer to Sect. 4 for more details about the pipeline.

6 Results

6.1 Overall Score

The overall score of each system is calculated as the average score of all the tasks. The results are shown in Fig. 3. The QA system outperforms the baseline system in terms of the overall score. The average score of the QA system is 69.78, while that of the baseline system is 64.96. However, the average wrong submissions and time taken by both systems are not significantly different. The average wrong submissions of the QA system is 0.42, while that of the baseline system is 0.48. The average time taken by the QA system is 77.17 s, while that of the baseline system is 74.78 s. The performance of each user is also shown in Fig. 4. The QA system outperforms the baseline system in terms of the overall score for 7 out of 10 users (except for users 5, 9, and 10).

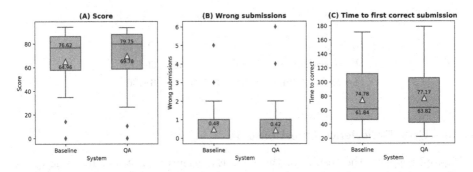

Fig. 3. (A) Overall score, (B) Time taken to answer the questions, and (C) Number of wrong submissions of the two systems.

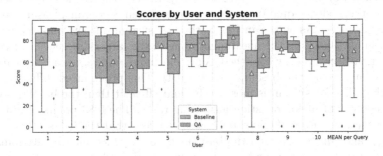

Fig. 4. Overall score of each user.

6.2 Importance of Experience

The results show that the QA system outperforms the baseline system in terms of the overall score. However, the performance of the QA system is not significantly better than the baseline system. This may be attributed to the fact that the participants had very little experience with lifelogging and question answering. To have a better understand of how the users perform with more experience, we analyse the average scores of the first system and the second system used by each user. The results are shown in Table 1. The average score of the first system used by each user is 66.67, while that of the second system used by each user is 68.06. This is expected as the users are more familiar with the tasks after using the first system. However, the average score of the first system used by the QA group (71.55) is higher than that of the baseline group (61.80). This indicates that the QA system is easier to use than the baseline system. Considering the second system only, the difference between two systems are not significant (68.12 for the baseline system and 68.00 for the QA system). The observed outcomes may be explained by the users getting familiar with the tasks after using the first system.

Table 1. Average scores for the first and second systems by each user.

System	Baseline	QA	Overall
First system only	61.80	**71.55**	66.67
Second system only	**68.12**	68.00	**68.06**

6.3   User Experience Questionnaire

Figure 5 displays the results of the User Experience Questionnaire. The questionnaire is designed to assess both pragmatic and hedonic aspects of system usability. The initial four questions measure the pragmatic quality of the system, focusing on its usefulness and efficiency. In contrast, the last four questions examine the hedonic quality, evaluating the system's overall pleasantness and user engagement. As shown in Fig. 5, the QA system outperforms the baseline system in all aspects in the questionnaire, with the larger difference observed in the pragmatic category, where the QA system shows an average advantage of 1.3 points compared to the baseline (1.7 vs. 0.4). This pronounced difference indicates that the QA system is more useful and efficient than the baseline system in the context of lifelog question answering tasks. The 0.83 points of difference in the hedonic category (1.5 vs. 0.67) also suggests that the QA system is more engaging and fun to use than the baseline system, which may be attributed to the QA system's intuitive and user-friendly nature, as discussed in the previous section.

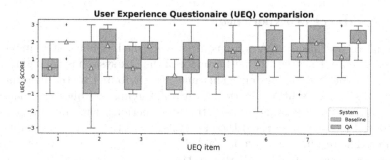

Fig. 5. Results of the User Experience Questionnaire.

7   Discussions and Conclusion

This paper presents a novel pipeline for integrating question answering capabilities into lifelog retrieval systems, which is based on the open-domain QA pipeline. By doing this, users can pose natural questions to the lifelogs and receive potential text answers. Our user study demonstrate the advantages of our QA system over the baseline approach in terms of overall scores and user satisfaction. Moreover, the results suggest that the QA system is a better option for new users.

In future works, deeper analysis on different question types is necessary to develop a well-rounded QA system. Furthermore, there are several ways to extend the QA pipeline, including result post-processing to improve the relevance of retrieved events, answer post-processing to re-rank the suggested answers, and answer highlighting to improve the user's confidence in the answers.

Acknowledgements. This work was conducted with the financial support of the Science Foundation Ireland Centre for Research Training in Digitally-Enhanced Reality (d-real) under Grant No. 18/CRT/6224. For the purpose of Open Access, the authors have applied a CC BY public copyright licence to any Author Accepted Manuscript version arising from this submission.

References

1. Alam, N., Graham, Y., Gurrin, C.: Memento 2.0: an improved lifelog search engine for LSC 2022. In: Proceedings of the 5th Annual on Lifelog Search Challenge, pp. 2–7 (2022)
2. Bain, M., Nagrani, A., Varol, G., Zisserman, A.: Frozen in time: a joint video and image encoder for end-to-end retrieval. In: IEEE International Conference on Computer Vision (2021)
3. Chang, C.C., Fu, M.H., Huang, H.H., Chen, H.H.: An interactive approach to integrating external textual knowledge for multimodal lifelog retrieval. In: Proceedings of the ACM Workshop on Lifelog Search Challenge, pp. 41–44 (2019)
4. Chen, D., Fisch, A., Weston, J., Bordes, A.: Reading wikipedia to answer open-domain questions. arXiv preprint arXiv:1704.00051 (2017)
5. Devlin, J., Chang, M.W., Lee, K., Toutanova, K.: BERT: pre-training of deep bidirectional transformers for language understanding. arXiv preprint arXiv:1810.04805 (2018)
6. Fabian Caba Heilbron, Victor Escorcia, B.G., Niebles, J.C.: ActivityNet: a large-scale video benchmark for human activity understanding. In: Proceedings of the IEEE Conference on Computer Vision and Pattern Recognition, pp. 961–970 (2015)
7. Gemmell, J., Bell, G., Lueder, R.: MyLifeBits: a personal database for everything. Commun. ACM **49**(1), 88–95 (2006)
8. Gurrin, C., et al.: Experiments in lifelog organisation and retrieval at NTCIR. In: Sakai, T., Oard, D.W., Kando, N. (eds.) Evaluating Information Retrieval and Access Tasks. TIRS, vol. 43, pp. 187–203. Springer, Singapore (2021). https://doi.org/10.1007/978-981-15-5554-1_13
9. Gurrin, C., et al.: Introduction to the fifth annual lifelog search challenge, LSC 2022. In: Proceedings of the International Conference on Multimedia Retrieval (ICMR 2022). ACM, Newark, NJ (2022)
10. Gurrin, C., et al.: Introduction to the sixth annual lifelog search challenge, LSC 2023. In: Proceedings of the International Conference on Multimedia Retrieval (ICMR 2023). ICMR 2023, New York (2023)
11. Izacard, G., Grave, E.: Leveraging passage retrieval with generative models for open domain question answering. arXiv preprint arXiv:2007.01282 (2020)
12. Karpukhin, V., et al.: Dense passage retrieval for open-domain question answering. arXiv preprint arXiv:2004.04906 (2020)

13. Laugwitz, B., Held, T., Schrepp, M.: Construction and evaluation of a user experience questionnaire. In: Holzinger, A. (ed.) USAB 2008. LNCS, vol. 5298, pp. 63–76. Springer, Heidelberg (2008). https://doi.org/10.1007/978-3-540-89350-9_6
14. Lee, J., Yun, S., Kim, H., Ko, M., Kang, J.: Ranking paragraphs for improving answer recall in open-domain question answering. arXiv preprint arXiv:1810.00494 (2018)
15. Lee, K., Chang, M.W., Toutanova, K.: Latent retrieval for weakly supervised open domain question answering. arXiv preprint arXiv:1906.00300 (2019)
16. Lewis, M., et al.: BART: denoising sequence-to-sequence pre-training for natural language generation, translation, and comprehension. arXiv preprint arXiv:1910.13461 (2019)
17. Lewis, P., et al.: Retrieval-augmented generation for knowledge-intensive NLP tasks. Adv. Neural. Inf. Process. Syst. **33**, 9459–9474 (2020)
18. Liu, Y., et al.: RoBERTa: a robustly optimized BERT pretraining approach. arXiv preprint arXiv:1907.11692 (2019)
19. Nguyen, T.N., et al.: Lifeseeker 3.0: an interactive lifelog search engine for LSC 2021. In: Proceedings of the 4th Annual on Lifelog Search Challenge, pp. 41–46 (2021)
20. Radford, A., et al.: Learning transferable visual models from natural language supervision. In: International Conference on Machine Learning, pp. 8748–8763. PMLR (2021)
21. Raffel, C., et al.: Exploring the limits of transfer learning with a unified text-to-text transformer. J. Mach. Learn. Res. **21**(1), 5485–5551 (2020)
22. Rajpurkar, P., Jia, R., Liang, P.: Know what you don't know: unanswerable questions for squad. arXiv preprint arXiv:1806.03822 (2018)
23. Spiess, F., Schuldt, H.: Multimodal interactive lifelog retrieval with vitrivr-VR. In: Proceedings of the 5th Annual on Lifelog Search Challenge, pp. 38–42 (2022)
24. Tran, L.-D., Ho, T.C., Pham, L.A., Nguyen, B., Gurrin, C., Zhou, L.: LLQA - lifelog question answering dataset. In: Þór Jónsson, B., et al. (eds.) MMM 2022. LNCS, vol. 13141, pp. 217–228. Springer, Cham (2022). https://doi.org/10.1007/978-3-030-98358-1_18
25. Tran, L.D., Nguyen, B., Zhou, L., Gurrin, C.: Myeachtra: event-based interactive lifelog retrieval system for LSC 2023. In: Proceedings of the 6th Annual ACM Lifelog Search Challenge, pp. 24–29. Association for Computing Machinery, New York (2023)
26. Tran, L.D., et al.: Comparing interactive retrieval approaches at the lifelog search challenge 2021. IEEE Access **11**, 30982–30995 (2023)
27. Tran, L.D., Nguyen, M.D., Nguyen, B., Lee, H., Zhou, L., Gurrin, C.: E-myscéal: embedding-based interactive lifelog retrieval system for LSC 2022. In: Proceedings of the 5th Annual on Lifelog Search Challenge, pp. 32–37. LSC 2022, Association for Computing Machinery, New York (2022)
28. Tran, L.D., Nguyen, M.D., Nguyen, B.T., Zhou, L.: Myscéal: a deeper analysis of an interactive lifelog search engine. Multimedia Tools Appl. **82**, 1–18 (2023)
29. Tran, Q.L., Tran, L.D., Nguyen, B., Gurrin, C.: MemoriEase: an interactive lifelog retrieval system for LSC 2023. In: Proceedings of the 6th Annual ACM Lifelog Search Challenge, pp. 30–35 (2023)
30. Wang, S., et al.: R 3: reinforced ranker-reader for open-domain question answering. In: Proceedings of the AAAI Conference on Artificial Intelligence, vol. 32 (2018)
31. Yang, A., Miech, A., Sivic, J., Laptev, I., Schmid, C.: Zero-shot video question answering via frozen bidirectional language models. arXiv preprint arXiv:2206.08155 (2022)

32. Yang, W., et al.: End-to-end open-domain question answering with BERTserini. arXiv preprint arXiv:1902.01718 (2019)

33. Yu, J., Wang, Z., Vasudevan, V., Yeung, L., Seyedhosseini, M., Wu, Y.: Coca: contrastive captioners are image-text foundation models. arXiv preprint arXiv:2205.01917 (2022)

34. Zhou, L., et al.: Overview of the NTCIR-16 lifelog-4 task. In: Proceedings of the 16th NTCIR Conference on Evaluation of Information Access Technologies, pp. 130–135. National Institute of Informatics (2022)

35. Zhu, F., Lei, W., Wang, C., Zheng, J., Poria, S., Chua, T.S.: Retrieving and reading: a comprehensive survey on open-domain question answering. arXiv preprint arXiv:2101.00774 (2021)

Event Recognition in Laparoscopic Gynecology Videos with Hybrid Transformers

Sahar Nasirihaghighi[1(✉)], Negin Ghamsarian[2], Heinrich Husslein[3], and Klaus Schoeffmann[1]

[1] Institute of Information Technology (ITEC), Klagenfurt University, Klagenfurt, Austria
`Sahar.Nasirihaghighi@aau.at`
[2] Center for AI in Medicine, University of Bern, Bern, Switzerland
[3] Department of Gynecology and Gynecological Oncology, Medical University Vienna, Vienna, Austria

Abstract. Analyzing laparoscopic surgery videos presents a complex and multifaceted challenge, with applications including surgical training, intra-operative surgical complication prediction, and post-operative surgical assessment. Identifying crucial events within these videos is a significant prerequisite in a majority of these applications. In this paper, we introduce a comprehensive dataset tailored for relevant event recognition in laparoscopic gynecology videos. Our dataset includes annotations for critical events associated with major intra-operative challenges and post-operative complications. To validate the precision of our annotations, we assess event recognition performance using several CNN-RNN architectures. Furthermore, we introduce and evaluate a hybrid transformer architecture coupled with a customized training-inference framework to recognize four specific events in laparoscopic surgery videos. Leveraging the Transformer networks, our proposed architecture harnesses inter-frame dependencies to counteract the adverse effects of relevant content occlusion, motion blur, and surgical scene variation, thus significantly enhancing event recognition accuracy. Moreover, we present a frame sampling strategy designed to manage variations in surgical scenes and the surgeons' skill level, resulting in event recognition with high temporal resolution. We empirically demonstrate the superiority of our proposed methodology in event recognition compared to conventional CNN-RNN architectures through a series of extensive experiments.

Keywords: Medical Video Analysis · Laparoscopic Gynecology · Transformers

1 Introduction

Laparoscopic surgery, an integral subset of minimally invasive procedures, has significantly transformed surgical practices by offering numerous benefits, such as reduced patient trauma, accelerated recovery time, and diminished postoperative discomfort [1]. Laparoscopic surgeries entail performing procedures through small incisions using a camera-equipped laparoscope [18,20]. The laparoscope is inserted through one of the trocars, and the camera provides a high-definition view of the abdominal structures

© The Author(s), under exclusive license to Springer Nature Switzerland AG 2024
S. Rudinac et al. (Eds.): MMM 2024, LNCS 14565, pp. 82–95, 2024.
https://doi.org/10.1007/978-3-031-56435-2_7

on a monitor. Laparoscopic videos capture a wealth of information critical for surgical evaluation, education, and research [17, 19]. Analyzing these videos encompasses diverse objectives, ranging from tool presence detection, surgical instrument tracking, and organ segmentation to surgical phase recognition and, crucially, surgical action or event recognition.

Surgical event recognition in laparoscopic gynecology surgery videos is a critical area of research within the field of computer-assisted intervention (CAI). Analyzing and recognizing surgical events in these videos holds great potential for improving surgical outcomes, enhancing training, and facilitating post-operative assessment [25]. However, even in the presence of advanced CAI systems [3], recognizing surgical events from laparoscopic surgery video recordings remains a formidable challenge. This complexity arises from the inherent variability in patient anatomy, the type of surgery, and surgeon skill levels [5], coupled with hindering factors such as the smoke presence within the surgical area [15], blurred lenses due to fog or blood, objects occlusions, and camera motion [30]. In this matter, we present and publicly release a comprehensive dataset for deep-learning-based event recognition in laparoscopic gynecology videos. Our dataset features four critical events, including Abdominal Access, Bleeding, Coagulation/Transaction, and Needle Passing. Since event recognition plays a prominent role in laparoscopic workflow recognition, the utilization of this dataset can effectively contribute to accelerating surgical training and improving surgical outcomes.

The combinations of CNNs-RNNs have proven to be a powerful technique in analyzing surgical videos, as they effectively leverage joint spatio-temporal information [6–9]. However, the evolution of event recognition has taken another significant turn with the recent adoption of transformer models [4, 29]. Originally designed for machine translation, the transformers have undergone a paradigm shift in their application to various time-series modeling challenges [26], including action recognition in laparoscopic surgery videos. The effectiveness of transformers stems from their remarkable self-attention mechanism, which enables capturing complex temporal relationships and dependencies within data sequences [2]. In this paper, we present a robust framework based on hybrid transformer models for relevant event recognition in laparoscopic gynecology videos to further enhance workflow analysis in this surgery.

This paper delivers the following key contributions:

- We introduce and evaluate a comprehensive dataset tailored for event recognition in gynecology laparoscopic videos. The video dataset, along with our customized annotations, has been publicly released.
  https://ftp.itec.aau.at/datasets/LapGyn6-Events/
- We propose a hybrid transformer-based model designed to identify significant events within laparoscopic gynecology surgery.
- We comprehensively evaluate the performance of the proposed model against several combinations of CNN-RNN frameworks.

The subsequent sections of this paper are structured as follows: Sect. 2 provides a comprehensive review of the relevant literature, highlighting distinctions from the present work. Section 3 presents a detailed explanation of the dataset, offering a thorough insight into its composition and characteristics. Section 4 delineates the proposed hybrid transformer-based framework for robust event recognition in laparoscopic gyne-cology surgery. The experimental setups are explained in Sect. 5, and the experimental results are presented in Sect. 6. Finally, Sect. 7 provides the conclusions drawn from the study.

2   Related Work

This section provides a comprehensive overview of the latest advancements in laparo-scopic surgical video analysis and action recognition within laparoscopy videos. Given the present study's focus on CNN-RNNs and Transformers, we aim to highlight methods that effectively utilize recurrent layers or Transformers for the analysis of laparoscopy videos.

Laparoscopic Surgery Video Analysis with CNN-RNNs: Tiwinanda et al. [28] intro-duce an architecture named EndoNet, crafted to concurrently perform phase recogni-tion and tool presence detection tasks within a multi-task framework. In [21], Namazi et al. introduced a surgical phase detection technique using deep learning system (SPD-DLS) to recognize surgical phases within laparoscopic videos. The approach employs a combination of the Convolutional Neural Network and Long Short-Term Memory model, which collectively analyzes both spatial and temporal information to detect sur-gical phases across video frames. Jin et al. proposed a CNN-LSTM for joint surgical phase recognition and tool presence detection [13]. While the CNN is responsible for extracting tool presence and spatial evidence of the surgical phase, the recurrent head is employed to model temporal dependencies between frame-wise feature maps. The study by Tobias et al. [3] presents a multi-stage temporal convolutional network (MS-TCN) tailored for surgical phase recognition. Termed TeCNO, derived from "Temporal Convolutional Networks for the Operating Room," this approach revolves around a sur-gical workflow recognition structure employing a two-stage TCN model, refining the extracted spatial features by establishing a comprehensive understanding of the cur-rent frame through an analysis of its preceding frames. In LapTool-Net by Namazi et al. [22], Gated Recurrent Unit (GRU) is employed to harness inter-frame correla-tions for multi-label instrument presence classification in laparoscopic videos. Golany et al. [10] suggest a multi-stage temporal convolution network for surgical phase recog-nition. In our previous work [23], we investigate action recognition in videos from laparoscopic gynecology with CNN-RNN model. In this paper, the detection of six actions-namely, abdominal access, anatomical grasping, knot pushing, needle pulling, needle pushing, and suction-from the gynecologic laparoscopy videos are explored by incorporating CNNs and cascaded bidirectional recurrent layers.

Action Recognition in Laparoscopy Videos with Transformers: Huang et al. [12] intro-duce a surgical action recognition model termed SA, leveraging the transformer's atten-tion mechanism. This encoder-decoder framework simultaneously classifies current and subsequent actions. The encoder network employs self-attention and cross-attention to grasp the intra-frame and inter-frame contextual relationships, respectively. Sub-sequently, the decoder predicts future and current actions, utilizing cross-attention to relate surgical actions. In [14], Kiyasseh et al. introduce a comprehensive surgical AI system (SAIS) capable of decoding various elements of intraoperative surgical activity from surgical videos. SAIS effectively segments surgical videos into distinct intraoperative activities. SAIS operates with two parallel streams processing separate input data types: RGB surgical videos and optical flow. Features from each frame are extracted using a pre-trained Vision Transformer (ViT) on ImageNet. These frame features then undergo transformation via a series of transformer encoders, yielding modality-specific video features. In another study, Shi et al. [27] present a novel method for surgical phase recognition, employing an attention-based spatial-temporal neural network. The approach leverages the attention mechanism within the spatial model, enabling adaptive integration of local features and global dependencies for enhanced spatial feature extraction. For temporal modeling, IndyLSTM and a non-local block are introduced to capture sequential frame features, establishing temporal dependencies among input clip frames. Sharghi et al. [11] introduce a dual-stage model comprising backbone and temporal components to address surgical activity recognition. The study highlights the effectiveness of the Swin-Transformer and BiGRU temporal model as a potent combination in achieving this goal.

3   Laparascopic Gynecology Dataset

We possess a dataset comprising 174 laparoscopic surgery videos extracted from a larger pool of over 600 gynecologic laparoscopic surgeries recorded with a resolution of 1920×1080 at the Medical University of Vienna. All videos are annotated by clinical experts into Abdominal Access, Grasping Anatomy, Transection, Coagulation, Knotting, Knot Pushing, Needle Passing, String Cutting, Suction, Irrigation, as well as the adverse event of Bleeding. While numerous actions or events are executed during laparoscopic surgery, we have specifically extracted four pivotal events of particular interest to physicians: (I) Abdominal Access, (II) Bleeding, (III) Coagulation and Transection, and (IV) Needle Passing. We detail these relevant events in the following.

Abdominal Access. The initial and one of the most critical steps in laparoscopic surgery involves achieving abdominal access by directly inserting a trocar with a blade into the abdominal cavity. This step holds particular significance, as it is the point where a substantial portion of potentially life-threatening complications, amounting to approximately $(40 - 50)\%$ of cases, can arise. These complications encompass critical issues like damage to major blood vessels, urinary tract injuries, and bowel damage [24]. While typically occurring as the first event, Abdominal Access can also take place at various points during the surgery.

Bleeding. Intraoperative hemorrhage stands as an adverse event potentially arising due to various factors, such as inadvertent cutting during the surgical process. Although some instances of bleeding during laparoscopic surgery are inevitable, they remain relatively rare and are proficiently addressed by the adept surgical team. Although laparoscopic surgery is specifically designed to minimize bleeding through the utilization of small incisions and precise techniques, bleeding events are probable. These events are paramount in laparoscopic surgery due to potentially affecting patient safety, surgical precision, and overall surgical outcomes.

Table 1. Visualization of relevant event annotations for ten representative laparoscopic videos from our dataset.

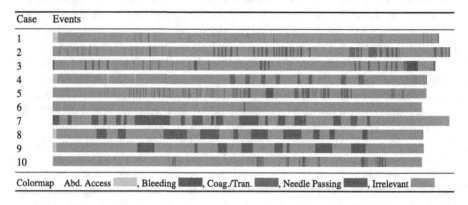

Table 2. Statistics corresponding to the four different relevant annotated cases in our dataset.

	Event	No. Cases	No. Segments	Segment Duration (sec)	Total duration (sec)
E1	Abd. Access	111	178	1 - 11	329.84
E2	Bleeding	41	81	2 - 108	1577.20
E3	Coag./Tran.	12	584	2 - 43	2929.54
E4	Needle passing	48	510	2 - 110	7036.43

Coagulation and Transection. These two events are frequently intertwined in laparoscopic surgery. Coagulation serves a crucial role in controlling bleeding at the transection site. This dual action ensures not only a clear surgical field but also a well-controlled surgical environment, allowing for precise removal of pathological tissues.

Needle Passing. This event encompasses the delicate movement of a needle into and out of the tissue, often conducted in tasks like suturing and tissue manipulation. Needle passing in laparoscopic surgery demands a particular skill set and a high level of precision, as it involves maneuvering the needle through tissue structures. Suturing, in particular, emerges as a cornerstone of laparoscopic surgery. Closing incisions with precise care can control bleeding to maintain a clear surgical field, repair damaged tissues and organs, and ultimately affect the overall success of the procedure. Beyond its technical significance, suturing in laparoscopy is a surgical skill assessment step [16].

Table 1 visualizes annotations for representative videos from our laparoscopic surgery dataset. Each row in the table corresponds to a distinct case of a laparoscopic surgery video. These cases exhibit varying durations, spanning from 11 min to approximately three hours. Within every individual case, the presence of distinct events, including Abdominal Access, Bleeding, Coagulation/Transection, and Needle Passing, along with irrelevant events in this study, is depicted using different colors. We should emphasize that not all laparoscopic videos contain all relevant events. The segments exhibit

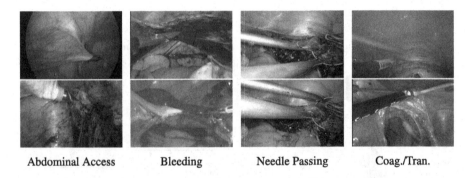

| Abdominal Access | Bleeding | Needle Passing | Coag./Tran. |

Fig. 1. Exemplar frames extracted from laparoscopic surgery videos.

varying durations, ranging from three seconds to over one minute in some cases. The details regarding the cases in each event are presented in Table 2. It's worth noting that the segments with durations of less than one second are excluded from the training set as they do not meet the criteria for our proposed *input dropout* strategy. Figure 1 depicts representative frames corresponding to these relevant events in laparoscopic gynecology surgery.

4   Proposed Approach

Within this section, we provide a comprehensive explanation of our proposed hybrid transformer model for the recognition of events in laparoscopic surgery videos. Additionally, we delineate our specialized training and inference framework.

4.1   Hybrid Transformer Model

The fusion of convolutional neural networks (CNNs) with transformers has emerged as a compelling approach for advancing surgical event recognition. This hybrid architecture capitalizes on the strengths of both components to achieve robust and accurate recognition of complex surgical events from video data. CNNs excel in extracting spatial features from individual frames, enabling them to capture essential visual information within each frame. These features contribute to the network's understanding of local patterns, textures, instruments, and anatomical structures present in laparoscopic videos. Integrating transformers into this architecture addresses the temporal dimension. Transformers' self-attention mechanisms enable them to learn global dependencies and long-range temporal associations within video sequences. By aggregating information from various frames, transformers capture the temporal context critical for recognizing the flow and progression of surgical events. Accordingly, we propose a hybrid transformer model for event recognition depicted in Fig. 2.

In our model, the CNN component functions as a proficient feature extractor, efficiently capturing intricate spatial details within each video frame. These spatial features

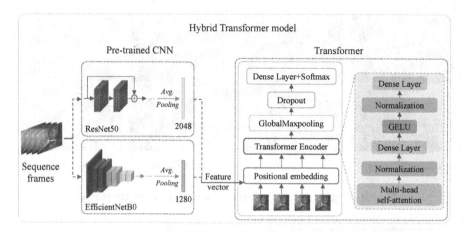

Fig. 2. Proposed hybrid transformer architecture.

provide a high-level representation of the input frames' spatial characteristics. Before passing the feature vectors through the transformer layers, we apply a positional embedding layer to the input frames. This layer adds positional information to each frame in the sequence to help the transformer model understand the temporal order of frames. The feature vectors with positional embeddings are then passed through the transformer encoder layers. These layers are responsible for capturing temporal relationships and dependencies between frames in the video sequence. They perform self-attention over the frames and apply feed-forward layers to process the sequence. After processing the video sequence through the transformer encoder layers, we apply global max-pooling to obtain a fixed-size representation of the entire video sequence. This operation aggregates information from all frames and summarizes it into a single vector. A dropout layer is added for regularization purposes to prevent overfitting to the irrelevant features in the input sequence. Finally, a dense layer followed by a Softmax activation function is used for event classification. This layer takes the fixed-size representation from the global max-pooling layer and produces class probabilities for video classification.

4.2   Training and Inference Framework

Our dataset comprises laparoscopic surgery videos ranging in duration from 10 min to 3 h, all encoded with a temporal resolution of 30 frames per second (see Sect. 5). The surgical events within these videos are temporally segmented and classified into four relevant events and the remaining as irrelevant events. These events span from less than a second to over a minute, depending on the specific surgical procedure and the patient's condition. Instead of directly feeding consecutive frames to the network, we adopt a dedicated frame sampling strategy termed *input dropout* to enhance the robustness of trained networks to the surgical content and speed variations. Concretely, we split all segments in each class into two-to-three-second video clips with a one-second overlap. We set the input sequence length for the network as ten frames. These frames are randomly selected from each 60-to-90-frame clip. This strategy maximizes

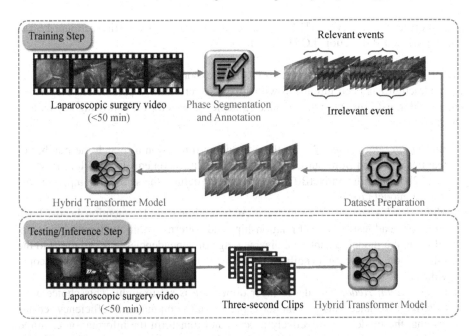

Fig. 3. Proposed training and inference framework for event recognition in gynecology laparoscopic videos.

variations in the combination of input frames presented to the network while preventing deviation of network parameters by unintended learning of the surgeon's speed. We also employ offline data augmentation techniques to negate the problem of class imbalance and increase the diversity of training data. Figure 3 provides a comprehensive depiction of our training and inference framework.

5   Experimental Setups

Data Augmentation: As mentioned in Sect. 4, we employ various data augmentation techniques to amplify the variety of training data accessible to the model, thereby mitigating overfitting risks and improving model generalization [9]. For each video segment in the dataset, We randomly select a set of spatial and non-spatial transformations, apply the same transformation to all frames within the segment, and re-encode the video segment. As spatial augmentations, we apply horizontal flipping with a probability of 0.5 and Gaussian blur with sigma = 5. Besides, the non-spatial transformations include gamma contrast adjustment with the gamma value being set to 0.5, brightness changes in the range of $[-0.3, 0.3]$, and saturation changes with a value of 1.5.

Alternative Methods: To validate the efficiency of our proposed hybrid transformer model, we proceed to train multiple instances of the CNN-RNN architecture. Concretely, we utilize the ResNet50 and EfficientNetB0 architectures as the backbone feature extraction component, complemented by four distinct variants of recurrent layers:

long short-term memory (LSTM), gated recurrent units (GRU), Bidirectional LSTM (BiLSTM), and Bidirectional GRU (BiGRU).

Neural Network Settings: In all networks, the backbone architecture employed for feature extraction has been initialized with ImageNet pre-trained parameters. Regarding our hybrid transformer model, we make strategic choices for the following key parameters:

1. *Embedding Dimension:* We set the embedding dimension to match the number of features extracted from our CNN model. This dimension equals 2048 for "ResNet" and 1280 for "EfficientNetB0". This parameter defines the size of the input vectors that the model processes.
2. *Multi-Head Attention:* We opt for a configuration with 16 attention heads. Each attention head learns distinct relationships and patterns within the input video data, allowing the model to capture a diverse range of dependencies and variations in the information, enhancing its ability to recognize intricate patterns within laparoscopic videos.
3. *Classification Head:* The dense layer dimension is set to eight. This choice aims to strike a balance between model capacity and computational efficiency, ensuring that the model can effectively process and transform the information captured by the attention mechanism. At the network's top, we sequentially add a Global-MaxPooling layer, a Dropout layer with a rate of 0.5, and a Dense layer followed by a Softmax activation.

In our alternative CNN-RNN models, the RNN component is implemented with a single recurrent layer consisting of 64 units. This design enables the model to incorporate high-level representations of features corresponding to individual frames, effectively capturing the subtle temporal dynamics within the video sequence. Following this, we implement dropout regularization with a rate of 0.4 to mitigate overfitting, followed by a Dense Layer composed of 64 units. Subsequently, we incorporate a global average pooling layer, followed by a dropout layer with a rate of 0.5. Ultimately, we conclude this architectural sequence with a Dense layer, employing a Softmax activation function.

Training Settings: The primary objective of this study is to identify and classify four critical events, including abdominal access, bleeding, needle passing, and coagulation/transection, within a diverse collection of gynecologic laparoscopic videos that include a multitude of events. To accomplish this task, we employ a binary classification strategy. In this approach, we divide the input videos into two distinct classes: the specific relevant event we seek to identify and all other events present in the video. This process is iteratively repeated for each of the four relevant events, resulting in the creation of four distinct binary classification models. In our analysis, the videos are resized to the resolution of 224×224 pixels. We train our hybrid transformer model on our video dataset using binary cross entropy loss function and the Adam optimizer with a learning rate of $\alpha = 0.001$. We assign 80% of the annotations to the training set and the remaining 20% to the test set.

Evaluation Metrics: For a comprehensive assessment of action recognition outcomes, we utilize four prominent evaluation metrics that are particularly well-suited for surgical action recognition. These metrics include accuracy, precision, and F1-score.

6 Experimental Results

Table 3 presents a comparative analysis of the proposed hybrid transformer's performance against various CNN-RNN architectures based on accuracy and F1-score. Notably, these results are derived from experiments conducted on both the original and the augmented datasets. Overall, the hybrid Transformer models exhibit higher average accuracy when applied to the augmented dataset compared to the original dataset. The experimental results reveal the superiority of the transformer model against the alternative CNN-RNN models when coupled with both ResNet50 and EfficientnetB0 networks. In particular, the combination of ResNet50 and the transformer model achieves the highest average accuracy (86.10%) and F1-score (86.03%), demonstrating its superior capability in identifying events within laparoscopic surgery videos. The second-best performing model comprises the EfficientNetB0 network with the Transformer,

Table 3. Quantitative comparisons between the proposed architecture and alternative methods based on accuracy and F1-score. The best results for each case are bolded.

		Original Dataset				
		Abd. Access	Needle Passing	Bleeding	Coag./Tran.	Average
Backbone	Head	Acc I F1 (%)	Acc I F1 (%)	Acc I F1 (%)	Acc I F1 (%)	Acc I F1 (%)
	LSTM	90.00 I 89.90	83.08 I 83.07	**77.43 I 77.40**	**88.03 I 87.94**	84.63 I 84.58
	GRU	91.25 I 91.18	85.54 I 85.48	74.78 I 74.72	86.32 I 86.11	84.47 I 84.37
ResNet50	BiLSTM	88.75 I 88.61	86.15 I 86.15	70.35 I 69.68	86.75 I 86.66	83.00 I 82.82
	BiGRU	87.50 I 87.30	84.31 I 84.30	73.45 I 73.15	87.39 I 87.33	83.16 I 83.02
	Transformer	**92.50 I 92.46**	**87.31 I 87.31**	74.34 I 73.67	86.97 I 86.85	**85.25 I 85.07**
	LSTM	86.25 I 86.20	85.85 I 85.68	75.22 I 74.94	86.11 I 85.92	83.35 I 83.18
	GRU	87.50 I 87.37	87.92 I 87.71	76.11 I 75.41	86.32 I 86.15	84.46 I 84.18
EfficientNetB0	BiLSTM	**88.75 I 88.71**	87.46 I 87.36	74.34 I 73.99	86.54 I 86.35	84.27 I 84.10
	BiGRU	87.50 I 87.43	83.62 I 83.23	**77.88 I 77.42**	**88.89 I 89.60**	84.47 I 84.42
	Transformer	86.25 I 86.25	**88.62 I 88.54**	75.66 I 75.58	88.46 I 88.37	**84.74 I 84.68**
		Augmented Dataset				
		Abd. Access	Needle Passing	Bleeding	Coag./Tran.	Average
Backbone	Head	Acc I F1 (%)	Acc I F1 (%)	Acc I F1 (%)	Acc I F1 (%)	Acc I F1 (%)
	LSTM	91.20 I 91.18	76.00 I 74.88	77.60 I 77.27	71.43 I 74.88	79.06 I 78.38
	GRU	90.00 I 89.90	75.69 I 75.05	74.80 I 74.80	67.31 I 63.92	76.95 I 75.92
ResNet50	BiLSTM	92.47 I 92.48	80.15 I 80.00	78.40 I 78.40	70.51 I 68.51	80.38 I 79.85
	BiGRU	91.25 I 91.18	77.69 I 77.02	78.00 I 77.84	71.67 I 70.10	79.65 I 79.03
	Transformer	**93.75 I 93.73**	**80.69 I 80.67**	**86.00 I 85.98**	**83.97 I 83.75**	**86.10 I 86.03**
	LSTM	**92.50 I 92.48**	**84.92 I 84.86**	80.40 I 80.26	71.15 I 69.80	82.24 I 81.85
	GRU	88.75 I 88.71	80.46 I 80.17	82.00 I 81.82	66.03 I 62.08	79.31 I 78.19
EfficientNetB0	BiLSTM	90.00 I 89.94	82.38 I 82.15	81.60 I 81.60	72.01 I 70.07	81.50 I 80.94
	BiGRU	91.25 I 91.22	83.31 I 83.02	83.60 I 83.46	72.65 I 71.45	82.70 I 82.32
	Transformer	88.75 I 88.73	82.85 I 82.83	**85.20 I 85.11**	**85.68 I 85.56**	**85.62 I 85.56**

achieving an average accuracy of 85.62% and an F1-score of 85.56%. It's important to highlight that the ResNet50-Transformer model achieves an impressive accuracy of 93.75% in correctly recognizing abdominal access events.

For further evaluation, we compared event precision across various network architectures, as illustrated in Fig. 4. In the case of ResNet50, it's evident that the abdominal access event stands out with remarkable precision when paired with the transformer model, achieving an impressive 94.44% accuracy. When considering the overall average precision, the ResNet50-Transformer combination emerges as the top performer. With the EfficientNetB0 backbone, the LSTM network demonstrates the highest precision for the abdominal access event, reaching an accuracy rate of 92.93%. Nevertheless, when examining the average precision across all events, the Transformer model consistently outperforms the other networks, showcasing its superiority in event recognition tasks.

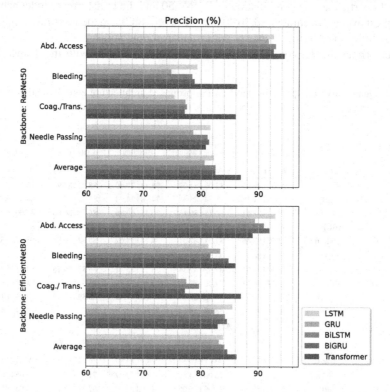

Fig. 4. Comparisons between the precision achieved by the proposed method versus alternative methods for relevant event recognition.

Figure 5 illustrates segments featuring two challenging events, Bleeding and Coagulation/Transection, recognized by our proposed ResNet50-Transformer model. The first segment consists of roughly 8 min from a laparoscopic surgery video, including various instances of bleeding events. Four randomly selected representative frames from

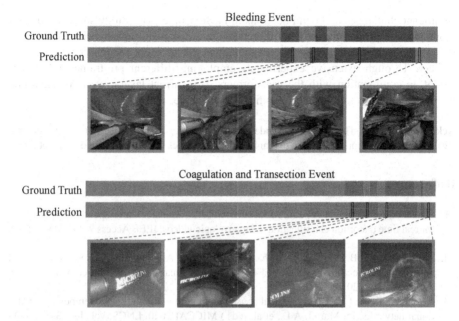

Fig. 5. Visualization of the proposed method's (ResNet50-Transformer) performance for a representative challenging segment of a laparoscopic video. The correct and incorrect predictions are shown with green and red borders, respectively. (Color figure online)

relevant or irrelevant events are depicted to showcase the challenging task of relevant event recognition in these videos. Remarkably, our model successfully identifies almost all bleeding events within the video. This achievement is significant given the inherent challenge of detecting bleeding, as it can occur during various surgical phases. The model's capability to recognize flowing blood, irrespective of the surgical instruments in use, underscores its robust performance. Of course, distinguishing between flowing and non-flowing blood can be quite complicated and lead to "false positive" detections, as demonstrated for the second visualized frame of the first sequence. Regarding the coagulation/transection events, it should be noted that laparoscopic videos can potentially contain smoky scenes, which increases the complexity of detecting this particular event. Despite these challenges, our model performs successfully in recognizing Coagulation/Transection events. However, in some instances, specific video clips annotated as irrelevant events are identified by the model as coagulation/transection event. This discrepancy can be attributed to the resemblance between these video clips and the actual coagulation/transection events.

7   Conclusion

Identifying relevant events in laparoscopic gynecology videos is a critical step in a majority of workflow analysis applications for this surgery. This paper has presented and evaluated a comprehensive dataset for recognizing four relevant events linked to significant challenges and complications in laparoscopic gynecology surgery. Furthermore, we have proposed a training/inference framework featuring a hybrid-transformer-based architecture tailored for the challenging task of event recognition within this surgery. Through extensive experiments employing different pre-trained CNN networks, in combination with the transformers or recurrent layers, we have identified the optimal architecture for our classification objective.

Acknowledgement. This work was funded by the FWF Austrian Science Fund under grant P 32010-N38. The authors would like to thank Daniela Stefanics for her help with data annotations.

References

1. Aldahoul, N., Karim, H.A., Tan, M.J.T., Fermin, J.L.: Transfer learning and decision fusion for real time distortion classification in laparoscopic videos. IEEE Access **9**, 115006–115018 (2021)
2. Bello, I., Zoph, B., Vaswani, A., Shlens, J., Le, Q.V.: Attention augmented convolutional networks. In: Proceedings of the IEEE/CVF international conference on computer vision, pp. 3286–3295 (2019)
3. Czempiel, T., et al.: TeCNO: surgical phase recognition with multi-stage temporal convolutional networks. In: Martel, A.L., et al. (eds.) MICCAI 2020. LNCS, vol. 12263, pp. 343–352. Springer, Cham (2020). https://doi.org/10.1007/978-3-030-59716-0_33
4. Dosovitskiy, A., et al.: An image is worth 16×16 words: transformers for image recognition at scale. arXiv preprint arXiv:2010.11929 (2020)
5. Funke, I., Mees, S.T., Weitz, J., Speidel, S.: Video-based surgical skill assessment using 3D convolutional neural networks. Int. J. Comput. Assist. Radiol. Surg. **14**, 1217–1225 (2019)
6. Ghamsarian, N.: Enabling relevance-based exploration of cataract videos. In: Proceedings of the 2020 International Conference on Multimedia Retrieval, pp. 378–382 (2020)
7. Ghamsarian, N., Amirpourazarian, H., Timmerer, C., Taschwer, M., Schöffmann, K.: Relevance-based compression of cataract surgery videos using convolutional neural networks. In: Proceedings of the 28th ACM International Conference on Multimedia, pp. 3577–3585 (2020)
8. Ghamsarian, N., Taschwer, M., Putzgruber-Adamitsch, D., Sarny, S., El-Shabrawi, Y., Schoeffmann, K.: LensID: A CNN-RNN-based framework towards lens irregularity detection in cataract surgery videos. In: de Bruijne, M., et al. (eds.) MICCAI 2021. LNCS, vol. 12908, pp. 76–86. Springer, Cham (2021). https://doi.org/10.1007/978-3-030-87237-3_8
9. Ghamsarian, N., Taschwer, M., Putzgruber-Adamitsch, D., Sarny, S., Schoeffmann, K.: Relevance detection in cataract surgery videos by spatio-temporal action localization. In: 2020 25th International Conference on Pattern Recognition (ICPR), pp. 10720–10727. IEEE (2021)
10. Golany, T., et al.: Artificial intelligence for phase recognition in complex laparoscopic cholecystectomy. Surg. Endosc. **36**(12), 9215–9223 (2022)
11. He, Z., Mottaghi, A., Sharghi, A., Jamal, M.A., Mohareri, O.: An empirical study on activity recognition in long surgical videos. In: Machine Learning for Health, pp. 356–372. PMLR (2022)
12. Huang, G.: Surgical action recognition and prediction with transformers. In: 2022 IEEE 2nd International Conference on Software Engineering and Artificial Intelligence (SEAI), pp. 36–40. IEEE (2022)
13. Jin, Y., et al.: Multi-task recurrent convolutional network with correlation loss for surgical video analysis. Med. Image Anal. **59**, 101572 (2020)

14. Kiyasseh, D., et al.: A vision transformer for decoding surgeon activity from surgical videos. Nat. Biomed. Eng. **7**, 780–796 (2023)
15. Leibetseder, A., Primus, M.J., Petscharnig, S., Schoeffmann, K.: Image-based smoke detection in laparoscopic videos. In: Cardoso, M.J., et al. (eds.) CARE/CLIP -2017. LNCS, vol. 10550, pp. 70–87. Springer, Cham (2017). https://doi.org/10.1007/978-3-319-67543-5_7
16. Lim, S., Ghosh, S., Niklewski, P., Roy, S.: Laparoscopic suturing as a barrier to broader adoption of laparoscopic surgery. J. Soc. Laparoendosc. Surg. **21**(3), e2017.00021 (2017)
17. Loukas, C.: Video content analysis of surgical procedures. Surg. Endosc. **32**, 553–568 (2018)
18. Loukas, C., Georgiou, E.: Smoke detection in endoscopic surgery videos: a first step towards retrieval of semantic events. Int. J. Med. Robot. Comput. Assist. Surg. **11**(1), 80–94 (2015)
19. Loukas, C., Varytimidis, C., Rapantzikos, K., Kanakis, M.A.: Keyframe extraction from laparoscopic videos based on visual saliency detection. Comput. Methods Programs Biomed. **165**, 13–23 (2018)
20. Lux, M., Marques, O., Schöffmann, K., Böszörmenyi, L., Lajtai, G.: A novel tool for summarization of arthroscopic videos. Multimedia Tools Appl. **46**, 521–544 (2010)
21. Namazi, B., Sankaranarayanan, G., Devarajan, V.: Automatic detection of surgical phases in laparoscopic videos. In: Proceedings on the International Conference in Artificial Intelligence (ICAI), pp. 124–130 (2018)
22. Namazi, B., Sankaranarayanan, G., Devarajan, V.: A contextual detector of surgical tools in laparoscopic videos using deep learning. Surg. Endosc. **36**, 679–688 (2022)
23. Nasirihaghighi, S., Ghamsarian, N., Stefanics, D., Schoeffmann, K., Husslein, H.: Action recognition in video recordings from gynecologic laparoscopy. In: 2023 IEEE 36th International Symposium on Computer-Based Medical Systems (CBMS), pp. 29–34 (2023)
24. Polat, M., Incebiyik, A., Tammo, O.: Abdominal access in laparoscopic surgery of obese patients: a novel abdominal access technique. Ann. Saudi Med. **43**(4), 236–242 (2023)
25. Schoeffmann, K., Del Fabro, M., Szkaliczki, T., Böszörmenyi, L., Keckstein, J.: Keyframe extraction in endoscopic video. Multimedia Tools Appl. **74**, 11187–11206 (2015)
26. Shi, C., Zheng, Y., Fey, A.M.: Recognition and prediction of surgical gestures and trajectories using transformer models in robot-assisted surgery. In: 2022 IEEE/RSJ International Conference on Intelligent Robots and Systems (IROS), pp. 8017–8024. IEEE (2022)
27. Shi, P., Zhao, Z., Liu, K., Li, F.: Attention-based spatial-temporal neural network for accurate phase recognition in minimally invasive surgery: feasibility and efficiency verification. J. Comput. Des. Eng. **9**(2), 406–416 (2022)
28. Twinanda, A.P., Shehata, S., Mutter, D., Marescaux, J., De Mathelin, M., Padoy, N.: EndoNet: a deep architecture for recognition tasks on laparoscopic videos. IEEE Trans. Med. Imaging **36**(1), 86–97 (2016)
29. Vaswani, A., et al.: Attention is all you need. In: Advances in Neural Information Processing Systems, vol. 30 (2017)
30. Wang, C., Cheikh, F.A., Kaaniche, M., Elle, O.J.: A smoke removal method for laparoscopic images. arXiv preprint arXiv:1803.08410 (2018)

GreenScreen: A Multimodal Dataset for Detecting Corporate Greenwashing in the Wild

Ujjwal Sharma(✉), Stevan Rudinac, Joris Demmers, Willemijn van Dolen, and Marcel Worring

University of Amsterdam, Amsterdam, The Netherlands
{u.sharma,s.rudinac,j.demmers,W.M.vanDolen,m.worring}@uva.nl

Abstract. Greenwashing, a form of deceptive marketing where organizations attempt to convince consumers that their offerings and operations are environmentally sound, can cause lasting damage to sustainability efforts by confusing consumers and eroding trust in genuine pro-sustainability actions. Nonetheless, capturing greenwashing "in the wild" remains challenging because greenwashed content frequently employs subliminal messaging through abstract semantic concepts that require subjective interpretation and contextualization within the context of the parent company's actual environmental performance. Moreover, this task typically presents itself as a weakly-supervised set-relevance problem, where the detection of greenwashing in individual media relies on utilizing supervisory signals available at the company level. To open up the task of detecting greenwashing in the wild to the wider multimedia retrieval community, we present a dataset that combines large-scale text and image collections, obtained from Twitter accounts for Fortune-1000 companies, with authoritative environmental risk scores on fine-grained issue categories like emissions, effluent discharge, resource usage, and greenhouse gas emissions. Furthermore, we offer a simple baseline method that uses state-of-the-art content encoding techniques to represent social media content and to understand the connection between content and its tendency for greenwashing.

Keywords: Greenwashing · Social multimedia · Multimedia modelling

1 Introduction

Over the past decade, there has been a noticeable surge in public awareness and sensitivity regarding climate change, its profound implications for the environment and human existence, as well as the significant role that consumers can play in mitigating its impact. This heightened consciousness is visible in the escalating demand for eco-friendly products and services. Nevertheless, instead of addressing the critical environmental concerns underlying the manufacturing and distribution of modern products and services, corporations have reacted to the environmentally conscious stance of consumers by employing a barrage of

© The Author(s), under exclusive license to Springer Nature Switzerland AG 2024
S. Rudinac et al. (Eds.): MMM 2024, LNCS 14565, pp. 96–109, 2024.
https://doi.org/10.1007/978-3-031-56435-2_8

misleading communication strategies. Such marketing endeavors deceive environmentally conscious consumers into believing that the firm's offerings and operations adhere to eco-friendly principles, and that their products meet the consumer's sustainability criteria [1,8,23]. These deceptive practices, collectively termed "greenwashing", pose an insidious threat to the public's trust in pro-sustainability initiatives, as they create confusion and mislead consumers into perceiving their chosen products or services as sustainable. Amidst this persistent dissemination of misinformation, social media has emerged as the latest and arguably the most influential platform for greenwashing. Large corporations are increasingly relying on dedicated social media teams to disseminate climate-related content. Given the prevalence of such aggressive marketing practices, consumers must possess the ability to discern such activities before their trust wanes.

Detecting greenwashing on social media presents significant challenges as this task is highly contextual. Posts from both ecologically-responsible and ecologically-irresponsible companies may share many similarities in the semantic content of their messaging, as both may claim to be environmentally conscious. However, greenwashing only becomes evident when the actions of the irresponsible company do not align with its words. From a technical standpoint, this implies that a potential solution to the task relies on the nature of both individual media (such as tweets, and images) and collective properties associated with the company as a whole. To capture greenwashing, the semantics of individual media elements, including images, tweets, videos, and other promotional materials, must be considered in the context of a company's real-world performance. This blend of individual and collective properties also extends to the supervisory signal – while the semantics of individual media are defined at the individual level, a company's performance on ecological issues falls into the collective category. Consequently, capturing greenwashing takes on the form of a multi-instance learning task that requires utilizing weaker group-level supervision to distinguish nuances within individual-level data. This process is further complicated by the fact that the signal within the mentioned weak supervision is diluted by the scale and noisy nature of individual media, much of which may not be related to greenwashing. An additional challenge stems from the high-level semantics inherent in the task. Since greenwashing relies on complex, subjectively interpreted, and polysemous visual and textual themes (such as images of lush forests or windmills), its detection is further complicated by the challenges associated with identifying high-level themes in media. An example of such complex themes is provided in Fig. 1.

To allow experimentation and analysis of corporate media streams as a social sensor for capturing common forms of greenwashing, and possibly, other deceptive marketing practices, we are publishing a large dataset that links the digital presence of Fortune 1000 – the largest American companies ranked by revenue, comprising of data posted on their Twitter[1] feeds to their performance on the

---

[1] Twitter is currently rebranding as X. However, for the remainder of this work, we will continue to refer to this platform by its former name – Twitter.

Fig. 1. Example posts that showcase greenwashing "in-the-wild". They exhibit how companies rely on complex, polysemous themes that hint at pleasantness or employ vague sustainability references that are difficult to quantify.

"Environmental" sub-components within the larger Environmental, Societal, and Governance (ESG) issues via ESG ratings. Specifically, our contributions are:

- We publish a dataset comprising nearly 1.1 million tweets and 650 thousand images from 423 Fortune-1000 companies, captured at the maximum level of detail via the official Twitter API.
- We provide annotations for these companies regarding seven ESG (Environmental, Social, and Governance) risk issues, which were collected from an ESG monitoring firm.
- We offer a straightforward baseline approach that demonstrates that a reasonable baseline performance for the task of predicting ESG risk ratings can be reached using state-of-the-art input image and text embedding models and give an indication of where improvements can be expected.

2 Related Work

2.1 Multimedia as a Social Sensor

Traditional benchmarks such as TREC [11], TRECVID [21], and MediaEval [15], were originally designed to tackle real-world challenges with significant societal consequences. These tasks frequently entailed working with real-world data and necessitated collaboration between communities within the problem domain and the computing community. Social media platforms have served as a powerful organic, nearly real-time social sensor for learning social themes such as online popularity [7], knowledge of fashion themes [16], themes in online public discourse during the COVID-19 pandemic [14], spread of rumors in online communities [24] and even predicting election results in multiple countries [3]. These studies capitalize on the organic and real-time nature of social media content to measure real-world effects, typically analyzing content at the individual instance level. While our proposed dataset also addresses a task that is becoming increasingly relevant, it goes beyond existing research by enabling research into problems that share a key characteristic with our task i.e. a hierarchical structure in which a collection of instances is grouped together within a bag or cluster, and the supervisory signal is only available at the bag level.

2.2 Measuring Sustainability

The assessment of corporate entities' performance on Environmental, Societal, and Governance (ESG) issues is a vital component of their "Social Responsibility". With modern consumers becoming increasingly discerning about the environmental impact of their purchases, corporations are placing greater emphasis on ensuring that their operations are viewed favorably from an ESG perspective. A substantial body of research in this area is concentrated within the Corporate Finance and Corporate Social Responsibility domains [2,9,10]. Corporations pay close attention to public perception because adverse ESG disclosures can result in lasting reputational damage and can impact stock prices [6,25]. To this effect, ESG ratings collected by financial monitoring firms assess the impact of environmental, social, and governance issues on a company and the company's effect on the outside world. These ratings are used as an investment tool that allows investors to numerically quantify the performance of firms on "material ESG issues" – significant factors that are relevant and have a measurable impact on a firm's financial performance. In essence, such ESG ratings represent the perceived degree of economic risk associated with investing in a company. Commonly available ESG ratings include MSCI ESG ratings[2], S&P Dow Jones[3] and Sustainalytics[4]. Our work represents an effort to unify the digital presence of companies with their actual sustainability efforts, enabling researchers to examine (1) whether their efforts are commensurate with their words and (2) discover unseen forms of greenwashing that may not be known yet.

---

[2] https://www.msci.com/our-solutions/esg-investing.

[3] https://www.spglobal.com/spdji/en/index-family/esg/.

[4] https://www.sustainalytics.com/esg-ratings.

3   The GreenScreen Dataset

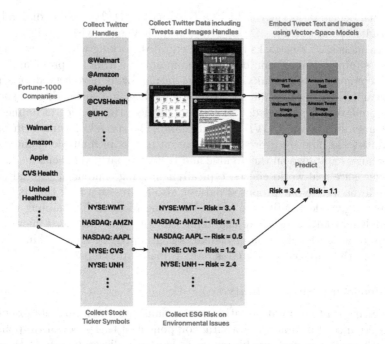

Fig. 2. An overview of this work. After compiling a list of Fortune 1000 companies, we proceed to manually acquire Twitter handles and download Tweet data for these entities. Furthermore, we gather financial data, such as Ticker symbols, to access ESG risk scores through a multi-step procedure. Finally, we formalize the task as a weakly-supervised learning problem where given a company (represented as a set of tweets/images), the goal is to predict the ESG risk score of a company on multiple ESG issues.

The GreenScreen dataset and associated resources associated can be found at https://uva-hva.gitlab.host/u.sharma/greenscreen

Research Challenges: Greenwashing poses a unique research challenge in that while the effect itself remains an entity-level phenomenon (greenwashing is a deliberate choice by a company), its means are individual media. Furthermore, to differentiate greenwashing from legitimate environmental efforts, the position of the company concerning sustainability measures must be known. Thus, we formulate this task as a set prediction task where given a set of images or tweet texts, the goal is to predict the risk rating for the parent company. At a high level, this setup is the traditional multiple-instance learning task where given a set of labeled bags comprising of unlabelled instances, the goal is to learn a prediction function that can predict the labels of unseen bags [4,13]. While complex formulations like [13] use attention modules to selectively re-project and attend to representations of the bag's contents, we discovered that such approaches do not scale well to the relatively large bag sizes for our task. Thus, we formulate a simple baseline that treats the task as a simple instance-level prediction task i.e. learning to predict the risk rating of a company from a single media (tweet text or image) instance. We detail the data collection and baseline models in the subsequent sections. A visual overview of this work is available in Fig. 2.

Twitter Data Collection: To establish a comprehensive multimodal dataset for assessing corporate greenwashing, we initiate the process by compiling a list of companies featured in the "Fortune 1000" - a compilation of the top 1,000 American companies based on revenue. These corporations are generally recognized as the most influential players in the American economic landscape and include both publicly traded and privately held entities. For each of these companies, we manually collected their official Twitter accounts (if they had one). In cases where the Twitter account for a major conglomerate holding company was either unavailable or significantly inactive, we substituted it with the Twitter account associated with their most prominent brand. Our initial compilation of Twitter accounts for Fortune 1000 companies yielded 605 companies with valid Twitter accounts. For each of these companies, we collected all tweets published between 1 January 2017 to 15 December 2022. We utilized the TWARC Python package [22] for collecting this data at the maximum level of detail to include tweet IDs, textual content, media links, contextual annotations, and author metadata available via the Twitter Search API. To ensure that we were collecting organic, stand-alone tweets from these companies, we modified the search query to drop tweets posted as a response to other tweets. After downloading these tweets, we systematically extracted media links associated with the collected tweets and downloaded all image assets associated with these links.

ESG Data Collection: The measurement of a company's performance on sustainability is a difficult task due to several reasons. First, quantifying a company's sustainability is challenging because formulating a precise numerical definition of sustainability is difficult; its definition relies on the specific sustainability risks faced by a company. Second, although the UN Sustainable Development Goals (SDGs) have been established as a blueprint for sustainability, assessing individual companies' adherence to these SDGs is complicated due to the broadly framed nature of the SDG statements. In the absence of concrete metrics to measure sustainability efforts, this task has fallen upon ESG risk ratings. These ratings measure a company's performance on a plethora of environmental, social, and governance issues considered "material" to the company and are used by investors and stakeholders to examine how sustainable are the operations, products, supply chain and overall impact of a company on the world. Specifically, these ratings compute a "risk score" that quantifies the risk to a company from multiple sub-issues within the ESG spectrum. This process consists of multiple steps. First, for each company, a "risk exposure" that quantifies the total risk to a company is computed. The computed exposure can then be subdivided into two sub-components.

1. Manageable Risk: The risk that a company can address through its initiatives.
2. Unmanageable Risk: Risk that cannot be reasonably managed through a company's internal efforts (e.g., risks arising from black-swan events).

The company's Manageable Risk can be further broken down into:

1. Managed Risk: The component of manageable risk actively being mitigated by the company through its initiatives.
2. Unmanaged Risk: The component of manageable risk that the company is not currently addressing.

The gap between manageable risk and managed risk is referred to as the "management gap" – the risk that can be addressed but is not being currently addressed by the company. In this work, we collected ESG risk scores from Sustainalytics, an independent ESG and corporate governance research firm whose ratings are available to research institutions via the Wharton Research Data Services program[5]. Sustainalytics defines the net risk score for an ESG issue as its "unmanaged risk" – the combination of the management gap (indicating the company's inability or lack of desire to address the issue) and unmanageable risk (reflecting risks inherent to the industry that cannot be reasonably managed). To generate aggregated ESG ratings, risk ratings for issues deemed significant to the company are weighted by their relative significance to the industry type and summed. Although this results in a single ESG score, it's important to observe that this work concerns itself with "greenwashing," which is primarily characterized by the misrepresentation of environmental concerns. Therefore, we steer clear of the combined ESG score and generate separate unmanaged risk scores for environmental issues. A visual overview of the risk computation process is provided in Fig. 3.

To collect ESG data, indices often use ticker symbols or unique, standardized identifiers that remain unmodified even in the case of corporate rebranding, as an identifier for the company. Getting ESG data for the set of companies with Twitter accounts was done by using fuzzy matching to link the publicly available company name often used on public accounts (for example, Apple) with its legal name (Apple Inc.). Once this was completed, we then linked this public name to its ticker symbol. For example, Apple Inc. is traded as "AAPL" on the NAS-DAQ stock exchange. Once the ticker symbol was available, we used a resource tool provided by WRDS to convert ticker symbols to standardized identifiers. We manually reviewed this file to remove any undesired entries (since the same ticker symbol may refer to different companies on different indices). The obtained company identifiers were then used to retrieve the company's risk score on material ESG issues. We manually verified that the downloaded scores corresponded with

---

[5] https://wrds-www.wharton.upenn.edu/pages/about/data-vendors/sustainalytics/.

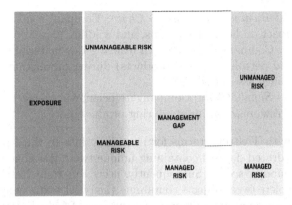

Fig. 3. An overview of the process for risk-assessment pipeline for companies within the Sustainalytics ESG framework.

the correct company and exchange and that there was no ambiguity. Finally, we retained the most current unmanaged risk score for each company on all environmental material issues. Since data for privately-held companies and companies without information on Sustainalytics could not be obtained, the set of companies under consideration was reduced from 605 to 423.

Dataset Structure: The resulting dataset comprises information on 423 companies annotated with ESG risk scores. Given that each company is represented as a collection of tweets (including text and images), this dataset assumes a bag-like structure commonly observed in Multiple Instance Learning tasks. In this structure, annotations are present at the bag level rather than for individual instances. This structure is generally more useful as greenwashing is a contextualized problem – posts about climate action do not automatically become greenwashing. Greenwashing is generally suspected in the presence of a large disparity between the stated goals of a company (as expressed in official media and promotions) and its real-world performance on sustainability-centric issues. Since real-world sustainability performance can only be measured for companies as a whole, retaining the bag-level structure of this task is generally more helpful.

The resulting dataset annotates companies with unmanaged risk scores on 8 material environmental issues. These issues include:

1. *Carbon - Own Operations:* Company's management of its own operational energy usage and greenhouse gas emissions.
2. *Carbon - Products and Services:* Energy efficiency and emissions associated with a company's products and services.
3. *Emissions, Effluents, and Waste:* Management of emissions from a company into the air, water, and land.
4. *Land Use and Biodiversity:* Company's management of its impact on land, ecosystems, and wildlife.

5. *Land Use and Biodiversity - Supply Chain:* Company's management of its suppliers' impact on land, ecosystems, and wildlife.
6. *Resource Use:* Company's efficiency in utilizing raw material inputs (excluding energy and petroleum-based products) during production, as well as its ability to manage associated risks.
7. *Resource Use - Supply Chain:* Company's efficiency in overseeing its suppliers' utilization of raw material inputs during production.

To comply with Twitter's terms for providing data to third parties, we are releasing a dehydrated dataset containing unique tweet IDs linked with companies and sustainability scores. Subsequently, users can "rehydrate" this dataset to generate complete tweet objects containing the tweet text, engagement statistics, geolocation information, and more. The process of "rehydration" can be executed using programs such as the previously mentioned TWARC or any Twitter library available to the user. Within the dataset, each company is annotated with unmanaged risk scores for the seven aforementioned environmental issues.

In total, the dataset comprises 1,120,058 tweets and 650,041 images. To eliminate the need for costly recomputation of embeddings, we include them within this dataset. The dataset contains image embeddings generated using state-of-the-art models, such as OpenAI's CLIP and a CLIP variant trained on the LAION-2B subset [18, 20, 26]. For tweet texts, we provide text embeddings computed with BERTweet-a publicly available, large-scale pretrained model for English tweets-and Sentence-BERT, a general-purpose sentence embedding model [17, 19]. In our experiments, these embeddings serve as static input embeddings for the primary task of predicting the target variable: unmanaged ESG risk.

To prepare this dataset for predictive tasks, we use a Group K-Fold splitting strategy where the dataset is split into companies rather than individual tweets or images to prevent information leakage between splits. Consequently, for instance-level tasks, all media from a company is limited to a single split. We retain 20% of the data for testing and the rest for training and validation splits.

4   Experiments and Baselines

4.1   Model Structure

The baseline task is to regress the unmanaged risk score for 7 distinct ESG issues from individual media such as tweet texts and images. We use a simple baseline design where static embeddings are mapped via a neural network with a single hidden layer to the target scalar unmanaged risk value. We utilize embeddings from state-of-the-art representation learning models to represent tweet images and text content. For image data, given that corporate content frequently includes visual concepts that may not be known beforehand and could fall outside the ImageNet-1000 subset, we opted to employ general-purpose models such as CLIP which have demonstrated an ability to zero-shot perform in many different tasks. Utilizing a CLIP model for image retrieval with SDG statements as search queries shows that CLIP embeddings inherently capture contextual information about the content semantics (cf. Fig. 4). To obtain vector-space representations for images, we employ OpenAI's CLIP, along with an additional CLIP variant trained on the LAION-2B dataset [5,12,18,20] as image-embedding models. For text, we make use of BERTweet, which is a BERT variant fine-tuned on a large-scale tweet collection, as well as Sentence-BERT [17,19] as tweet text-embedding model. For each model, we employ a locked-tuning approach, similar to the method described in [27]. In this approach, both image and text representations remain frozen, and an MLP-based head is trained to learn a mapping from the fixed representation to a scalar value denoting the unmanaged risk score for the media on an ESG issue. Since no gradients are computed for the image or text embedding models, embeddings can be precomputed allowing a significant speedup in training speed.

Fig. 4. To assess the effectiveness of CLIP embeddings as input encodings, we evaluated their ability to retrieve relevant images based on an SDG search query. Each block begins with the SDG query, and below it, you can find the top 5 results.

4.2   Model Training and Evaluation

We conducted the recommended image preprocessing before feeding images into CLIP. Similarly, for tweets, we performed the tweet normalization recommended in [17] where we replaced all user account mentions with @USER and all HTTP URLs with HTTPURL before feeding them into text-embedding models. We employed a straightforward Multilayer Perceptron (MLP) with just one hidden layer as our model. The purpose of the hidden layer was to act as a bottleneck, decreasing the dimensionality of the input data to one-eighth of its original size. This reduction was then followed by a final projection to a real-valued scalar, representing the predicted risk. Since the baseline is a regression task, we operationalize the standard regression measures – Mean Absolute Error (MAE), Mean Squared Error (MSE), and Root Mean Squared Error (RMSE) as evaluation measures. Additionally, we initialize Kendall's rank correlation coefficient (τ) as an additional measure to check if the risk score predicted by the model exhibits a correlation with the ground-truth unmanaged ESG risk rating. All models were trained for 200 epochs using a batch size of 256, stochastic gradient descent (with momentum set to 0.9) as the optimizer, and a learning rate of 5×10^{-5} with early stopping enabled. The best model had the lowest MAE score on the validation set. Since our instance-level formulation assumes that media instances within a bag carry the same risk label as the bag (company) itself, we calculate all measures at both the instance and bag levels. Instance-level statistics do not consider the bag-level structure of the problem and compute measures assuming that the task is purely an instance-level regression task. Bag-level measures aggregate individual risk predictions for instances within a bag, combining them into a single risk score using mean-pooling.

5   Results

Table 1. Results for models trained on image inputs embedded with OpenAI CLIP. Values under the IL section denote instance-level measures (computed by calculating the measure between individual instances), while BL denotes bag-level measures (computed by aggregating predictions for instances within a bag into a single prediction using mean-pooling and then calculating the measure).

Issue	MAE		MSE		RMSE		Kendall's τ	
Measure Type	IL	BL	IL	BL	IL	BL	IL	BL
Carbon (Own Operations)	1.20	1.16	2.55	2.29	1.59	1.51	0.38	0.57
Carbon (Products & Services)	0.48	0.60	1.05	1.35	1.02	1.16	0.26	0.42
Emissions, Effluents and Waste	0.90	1.04	1.82	1.90	1.35	1.38	0.35	0.50
Land Use & Biodiversity	0.23	0.22	0.31	0.26	0.55	0.51	0.30	0.40
Land Use & Biodiversity (Supply Chain)	0.14	0.21	0.13	0.21	0.36	0.45	0.20	0.39
Resource Use	0.12	0.18	0.09	0.19	0.30	0.43	0.12	0.30
Resource Use (Supply Chain)	0.71	0.85	1.04	1.15	1.02	1.07	0.30	0.51

In our testing, we observed that OpenAI's CLIP model and the Sentence-BERT models outperformed CLIP-LAION and BERTweet respectively by small margins. The results for the baseline models are presented in Tables 1 and 2. These results indicate that prediction errors are reasonably small for both instance and

Table 2. Results for models trained on tweet text inputs embedded with Sentence-BERT. Values under the IL section denote instance-level measures (computed by calculating the measure between individual instances), while BL denotes bag-level measures (computed by aggregating predictions for instances within a bag into a single prediction using mean-pooling and then calculating the measure).

Issue	MAE		MSE		RMSE		Kendall's τ	
Measure Type	IL	BL	IL	BL	IL	BL	IL	BL
Carbon (Own Operations)	1.26	1.18	2.93	2.53	1.71	1.59	0.42	0.53
Carbon (Products & Services)	0.44	0.60	0.82	1.16	0.90	1.07	0.23	0.40
Emissions, Effluents and Waste	0.95	1.03	1.98	1.92	1.40	1.38	0.41	0.52
Land Use & Biodiversity	0.27	0.22	0.38	0.26	0.61	0.51	0.40	0.42
Land Use & Biodiversity (Supply Chain)	0.14	0.22	0.12	0.23	0.35	0.48	0.26	0.43
Resource Use	0.77	0.85	1.16	1.21	1.08	1.10	0.32	0.48
Resource Use (Supply Chain)	0.16	0.23	0.09	0.24	0.30	0.49	0.04	0.16

bag-level measures. Furthermore, we also observed moderate rank-correlations between model predictions and the ground-truth unmanaged ESG risk. We also observed that in most cases, instance-level errors are much larger than bag-level errors. This can be partly explained by the fact that the dataset has staggeringly different statistics at bag and instance-levels. At the bag-level, the entire dataset consists of 423 companies. However, at the instance-level, the count of individual tweet texts is 1,120,058 and the count of images is 650,041. Consequently, more precise measures for evaluating bag-level predictions may be needed.

6   Conclusions

Greenwashing and deceptive marketing represent complex socio-economic issues. While a long-term, comprehensive solution to this problem may necessitate the involvement of multiple stakeholders, GreenScreen offers an initial step in that direction. This dataset comprises nearly 650 thousand images, 1.1 million tweets, and ESG ratings for 423 companies, providing a starting point for addressing false environmental marketing and ensuring that corporations remain accountable for their communications. Beyond the uniqueness of the dataset, it presents an interesting task for the community, as it requires identifying greenwashing within individual media content through indirect supervisory signals provided by the companies that post such media. Furthermore, this task also presents challenges in terms of semantic relevance within extensive collections, as company tweets on Twitter encompass a wide range of topics, not all of which are related to greenwashing. We believe that introducing this innovative task to the multimedia research community can foster exploration of this issue, ultimately leading to a world where consumers can actively recognize and address potential deception on online platforms.

References

1. Bonneuil, C., Choquet, P.L., Franta, B.: Early warnings and emerging account-ability: total's responses to global warming, 1971–2021. Glob. Environ. Chang. **71**, 102386 (2021)
2. Broadstock, D.C., Chan, K., Cheng, L.T., Wang, X.: The role of ESG performance during times of financial crisis: evidence from COVID-19 in china. Financ. Res. Lett. **38**, 101716 (2021)
3. Cai, H., Yang, Y., Li, X., Huang, Z.: What are popular: exploring twitter features for event detection, tracking and visualization. In: Proceedings of the 23rd ACM International Conference on Multimedia, pp. 89–98. MM 2015, Association for Computing Machinery, New York (2015)
4. Carbonneau, M.A., Cheplygina, V., Granger, E., Gagnon, G.: Multiple instance learning: a survey of problem characteristics and applications. Pattern Recogn. **77**, 329–353 (2018)
5. Cherti, M., et al.: Reproducible scaling laws for contrastive language-image learn-ing. In: Proceedings of the IEEE/CVF Conference on Computer Vision and Pattern Recognition, pp. 2818–2829 (2023)
6. Deng, X., Cheng, X.: Can ESG indices improve the enterprises' stock market performance?-an empirical study from China. Sustainability **11**(17), 4765 (2019)
7. Ding, K., Wang, R., Wang, S.: Social media popularity prediction: a multiple fea-ture fusion approach with deep neural networks. In: Proceedings of the 27th ACM International Conference on Multimedia, pp. 2682–2686 (2019)
8. Dunlap, R.E., McCright, A.M., et al.: Organized climate change denial. Oxford Handb. Clim. Change Soc. **1**, 144–160 (2011)
9. Gillan, S.L., Koch, A., Starks, L.T.: Firms and social responsibility: a review of ESG and CSR research in corporate finance. J. Corp. Finan. **66**, 101889 (2021)
10. Halbritter, G., Dorfleitner, G.: The wages of social responsibility - where are they? A critical review of ESG investing. Rev. Financ. Econ. **26**, 25–35 (2015)
11. Harman, D.: Overview of the first TREC conference. In: Proceedings of the 16th Annual International ACM SIGIR Conference on Research and Development in Information Retrieval, pp. 36–47 (1993)
12. Ilharco, G., et al.: Openclip (2021)
13. Ilse, M., Tomczak, J., Welling, M.: Attention-based deep multiple instance learning. In: International Conference on Machine Learning, pp. 2127–2136. PMLR (2018)
14. Kumar, A., Garg, G.: Sentiment analysis of multimodal twitter data. Multimedia Tools Appl. **78**, 24103–24119 (2019)
15. Larson, M., et al.: Automatic tagging and geotagging in video collections and com-munities. In: Proceedings of the 1st ACM International Conference on Multimedia Retrieval. ICMR 2011, Association for Computing Machinery, New York (2011)
16. Ma, Y., Yang, X., Liao, L., Cao, Y., Chua, T.S.: Who, where, and what to wear? Extracting fashion knowledge from social media. In: Proceedings of the 27th ACM International Conference on Multimedia, pp. 257–265. MM 2019, Association for Computing Machinery, New York (2019)
17. Nguyen, D.Q., Vu, T., Nguyen, A.T.: BERTweet: a pre-trained language model for English tweets. In: Proceedings of the 2020 Conference on Empirical Methods in Natural Language Processing: System Demonstrations, pp. 9–14 (2020)
18. Radford, A., et al.: Learning transferable visual models from natural language supervision. In: International Conference on Machine Learning, pp. 8748–8763. PMLR (2021)

19. Reimers, N., Gurevych, I.: Sentence-BERT: sentence embeddings using siamese BERT-networks. In: Proceedings of the 2019 Conference on Empirical Methods in Natural Language Processing. Association for Computational Linguistics (2019). https://arxiv.org/abs/1908.10084

20. Schuhmann, C., et al.: LAION-5b: an open large-scale dataset for training next generation image-text models. In: Thirty-sixth Conference on Neural Information Processing Systems Datasets and Benchmarks Track (2022). https://openreview.net/forum?id=M3Y74vmsMcY

21. Smeaton, A.F., Over, P., Kraaij, W.: Evaluation campaigns and TRECVid. In: Proceedings of the 8th ACM International Workshop on Multimedia Information Retrieval, pp. 321–330 (2006)

22. Summers, E., et al.: Docnow/twarc: v2.13.0 (2022). https://doi.org/10.5281/zenodo.7484102

23. Supran, G., Oreskes, N.: Assessing ExxonMobil's climate change communications (1977–2014). Environ. Res. Lett. **12**(8), 084019 (2017)

24. Tian, L., Zhang, X., Wang, Y., Liu, H.: Early detection of rumours on twitter via stance transfer learning. In: Jose, J.M., et al. (eds.) ECIR 2020. LNCS, vol. 12035, pp. 575–588. Springer, Cham (2020). https://doi.org/10.1007/978-3-030-45439-5_38

25. Wong, J.B., Zhang, Q.: Stock market reactions to adverse ESG disclosure via media channels. Br. Account. Rev. **54**(1), 101045 (2022)

26. Zhai, X., Kolesnikov, A., Houlsby, N., Beyer, L.: Scaling vision transformers. In: Proceedings of the IEEE/CVF Conference on Computer Vision and Pattern Recognition, pp. 12104–12113 (2022)

27. Zhai, X., et al.: Lit: zero-shot transfer with locked-image text tuning. In: Proceedings of the IEEE/CVF Conference on Computer Vision and Pattern Recognition, pp. 18123–18133 (2022)

Author Index

S. Rudinac et al. (Eds.): MMM 2024, LNCS 14565, p. 111, 2024.
https://doi.org/10.1007/978-3-031-56435-2

Printed in the United States
by Baker & Taylor Publisher Services